CONCEPTS AND TECHNIQUES IN MODERN (

LINEAR REGRESSION IN GEOGRAPHY

by

Rob Ferguson

(University of Stirling)

CONTENTS

Acknowledgements

To Joyce Bell and Keith Scurr of the Geography Department at Hull University, for typing the text and drawing the diagrams; John Pethick for many discussions; and Jim Thompson (Hull) and Pete Taylor (Newcastle) for technical comments.

LINEAR REGRESSION IN GEOGRAPHY

I INTRODUCTION

(i) Relationships between variables

Most if not all systems studied by geographers involve variable quantities that can be measured numerically, from the distances travelled by Sunday trippers to the hydraulic characteristics of proglacial streams. Comparison of maps or time series graphs often shows that differences in one variable, in space or over time, appear to be associated with differences in other variables. For example both Sunday trips and proglacial streamflow may vary according to temperature. Sometimes the apparent association is entirely a matter of chance, and in other cases two variables may amount to alternative definitions of the same thing, but frequently we suspect cause and effect are at work. A relationship of this kind can be important in three ways. It may confirm or refute theoretical notions about cause-effect processes in the system under study. Alternatively it may highlight something not considered by existing theory and thus stimulate new ideas. Thirdly, it may be useful for prediction of the response to future changes in conditions, or of the present state of affairs in places where direct measurement is inconvenient or impossible.

Whether we are testing, generalising, or predicting, an objective method for summarising the form and strength of apparent associations is useful. Occasionally two or more variables are linked by a simple mathematical equation, as in the 'laws' of physics. But most geographical relationships are only broad trends, with individual cases departing from the norm because of unique local circumstances and differences in other relevant factors. The statistical techniques used to pick out general tendencies of this kind are known as regression methods. It is not surprising that they are very widely used in geography.

(ii) Aims and prerequisites

Regression analysis is a major branch of mathematical statistics and is used throughout the social and environmental sciences as well as in many branches of industry, business, and government. As a result computer programs are widely available, there is a vast literature on the subject, and many specialised variants of the basic techniques have been devised. Attention is focused here on the simplest and commonest version, linear regression. It is unrealistic to expect all users of regression methods, including those who make judgments on the basis of other peoples' analyses, to understand in detail how these tools work. But pitfalls surround the unwary user who has no idea what is involved. I have therefore tried to explain how linear regression works, what kind of underlying model it imposes on reality, and what can be done to check the assumptions made, as well as explaining how to interpret the results. The treatment is introductory with worked examples, graphical illustrations, and no mathematics beyond simple algebra. More advanced treatments, usually involving calculus and matrix algebra, can be found in a wide range of textbooks of which a few are listed in the bibliography. The statistical level is also kept low and familiarity with simple regression is not essential. The reader is however assumed to have taken an introductory course covering elementary descriptive statistics, including the correlation coeffecient, and significance testing. Finally, the notation adopted here is widely used and fairly obvious but the reader is warned that other authors may use alternative symbols for the same things and the same symbols for different things.

(iii) Averages, variances, and correlations

Most explanations of regression analysis obtain and present the main results in terms of sums of squares of values of each variable and sums of products of different variables. Here we work entirely in terms of the three basic descriptive statistics: means, standard deviations, and correlations. There are several advantages. These statistics should be familiar to all geographers. In most applications they will have been calculated and inspected for their own interest before any further analysis is comtemplated. Their numerical values are immediately interpretable, and also generally small which reduces the danger of roundoff errors in machine computation or shifted decimal places and the like in hand calculations. And their use avoids a profusion of summation signs hereafter.

Means, standard deviations, and correlations are all defined in terms of averages, and the operation of averaging plays a vital role later on. It is represented here by putting a bar over the quantity to be averaged. The simplest case is $\overline{X}$, which stands for the arithmetic mean $\Sigma X/n$ of n values of a numerical variable X. More complicated expressions can be averaged using three simple rules: (1) multiply out any brackets; (2) the average of a sum is a sum of separate averages; (3) constants can be moved outside averages. For example, if k is a constant the mean of k(X + Y) is

$$\overline{k(X + Y)} = \overline{kX + kY} = \overline{kX} + \overline{kY} = k\overline{X} + k\overline{Y}$$

beyond which it cannot be simplified.

For a variable to merit the name its individual values must deviate from their mean. Deviations are represented here by lower case letters, for example x for the deviations $X - \overline{X}$ of X from its mean $\overline{X}$. The usual measure of variability is the standard deviation, denoted here by s with a subscript to show which variable it refers to. Thus s_X is the standard deviation of X. Its square is the variance, defined as the mean squared deviation:

$$s_X^2 = \text{variance of } X = \overline{x^2} = \overline{(X - \overline{X})^2} \, .$$

Application of the averaging rules converts this to

$$s_X^2 = \overline{X^2 - 2\overline{X}X + \overline{X}^2} = \overline{X^2} - (\overline{X})^2$$

which is one version of the familiar short-cut formula used to calculate variances and standard deviations. But for our purposes it is more important that if a term like $\overline{x^2}$ occurs in an average it can be recognised as simply the variance or squared standard deviation of the variable concerned.

Regression analysis is about relationships between variables and the contributions of different factors to overall variability. It is therefore useful to be able to split into components the variance of quantities like X + Y. Rules (1) to (3) show that

$$\text{variance of } X + Y = \overline{(x+y)^2} = \overline{x^2 + 2xy + y^2}$$

$$= \overline{x^2} + 2\overline{xy} + \overline{y^2}.$$

The outside terms are simply the variances s_x^2, s_y^2 of X and Y, but the middle term involves a new kind of average, the mean product $\overline{xy}$ of corresponding X and Y deviations. This is called the covariance of, or between, X and Y. It can be shown that its maximum possible value is the product of the standard deviations of the two variables, when corresponding X and Y deviations are all in exactly the same proportion. The familiar correlation coefficient, r, is simply the ratio of actual to maximum possible covariance:

$$\text{correlation between X and Y} = r_{xy} = \overline{xy}/s_x s_y.$$

More familiar formulae for r can be obtained using the averaging rules, but the important point for our purposes is that the covariance of two variables can be rewritten as their correlation times the product of their standard deviations. So the variance of X + Y above reduces to

$$\overline{(x+y)^2} = s_x^2 + 2r_{xy}s_x s_y + s_y^2$$

i.e. a sum of separate contributions plus a joint one that only disappears if X and Y are uncorrelated.

The correlation coefficient is of course a measure of association. If high X tends to go with high Y and low with low then r is positive, approaching its limit of +1 the stronger the tendency. If high X goes with low Y and vice versa r is negative, while if there is no pattern one way or the other r is close to zero. (see Fig. 1). A correlation can be calculated between any pair of numerical variables, including those whose values are only ranked (when r = Spearman's rank correlation) or even binary (when $nr^2 = \chi^2$ for the 2x2 frequency table of values). Relationships between such variables do not possess form, apart from being positive or negative, so regression analysis is normally restricted to interval or ratio scale data. Some exceptions are mentioned in the final chapter, but otherwise we will be concerned with relationships between variables with a more or less continuous range of possible values.

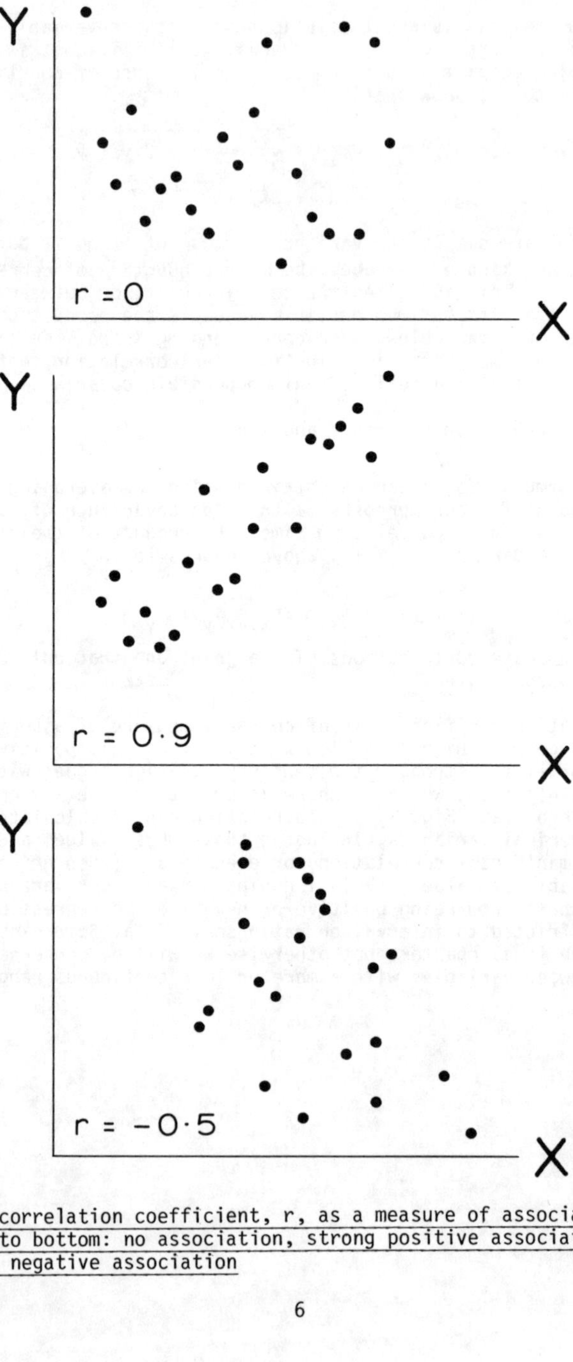

Fig. 1 The correlation coefficient, r, as a measure of association. Top to bottom: no association, strong positive association, weak negative association

II SIMPLE REGRESSION

(i) The least squares trend

The observed relationship between two numerical variables X and Y can be represented exactly by a scatter diagram or graph of Y values against X values, with one point for each pair of values (each site or area in a spatial study, person or household in a social study, time or time interval in a historical study, and so on). As an illustration Figure 2 shows the relationship between long term average precipitation (Y) and height above sea level (X) at twenty rain gauges in a west-east transect across southern Scotland (the data are in Table 1).

If any association can be discerned in such a scatter diagram it can be summarised with greater or lesser accuracy by the equation of a trend curve passing through the cloud of points. Any type of curve can be chosen but the simplest possibility, perfectly adequate in this example, is the straight line

$$\hat{Y} = a + bX \tag{1}$$

which predicts the trend value $\hat{Y}$ of Y for any given X value. The constants a and b are respectively the intercept and slope of the trend line: that is, $\hat{Y} = a$ when $X = 0$ and $\hat{Y}$ increases by b units (possibly negative) when X increases by one. In this case it is clear that rainfall increases rather than decreases with elevation so b is positive.

Simple regression is the process of choosing appropriate values of the regression coefficients a and b, or in effect choosing the particular straight line that best describes the trend of the data. This is generally done by the method of least squares, in which the coefficients are chosen to minimise the sum of squares (or equivalently the mean square) of the residuals or differences between observed and predicted Y values for each observed X value. Each residual is given by $e = Y - \hat{Y}$ and can be viewed as the vertical distance of a data point from the trend line as shown in Fig. 2.

Minimisation of the residual scatter involves elementary calculus and only the results are given here (for a proof see any advanced text). For a given slope, b, changes in the intercept, a, shift the trend line up or down. The residual scatter is found to be least when

$$a = \overline{Y} - b\overline{X} \tag{2a}$$

which implies that the best-fit trend line runs through the mean or centre of gravity of the data points, and that $\overline{e} = 0$ or residuals above and below the line cancel out. This fixes the general level of the line, but its tilt depends on b. It turns out that the residual scatter is least for a slope of

$$b = r\, s_y/s_x. \tag{2b}$$

This slope is measured in units of Y per unit of X, so its numerical value depends on the units chosen.

Table 1: Average precipitation and elevation across southern Scotland

Site No.	elevation (m above OD)	rainfall (mm/yr)	Site No.	elevation (m above OD)	rainfall (mm/yr)
1	240	1720	11	140	1460
2	430	2320	12	540	1860
3	420	2050	13	280	1670
4	470	1870	14	240	1580
5	300	1690	15	200	1490
6	150	1250	16	210	1420
7	520	2130	17	160	900
8	460	2090	18	270	1250
9	300	1730	19	320	1170
10	410	2040	20	230	1170

Source: British Rainfall (HMSO), selected raingauges between national grid lines 600 and 601 km N. Sites are in west-east order

Table 2: Residuals from rainfall-elevation trend

Site No.	residual (mm/yr)	distance E (km from W. coast)	Site No.	residual (mm/yr)	distance E (km from W. coast)
1	254	37	11	232	86
2	402	43	12	-319	97
3	156	48	13	109	100
4	-143	49	14	114	103
5	81	52	15	119	104
6	-2	59	16	25	114
7	-2	73	17	-376	138
8	101	75	18	-287	151
9	121	76	19	-486	153
10	170	77	20	-272	154

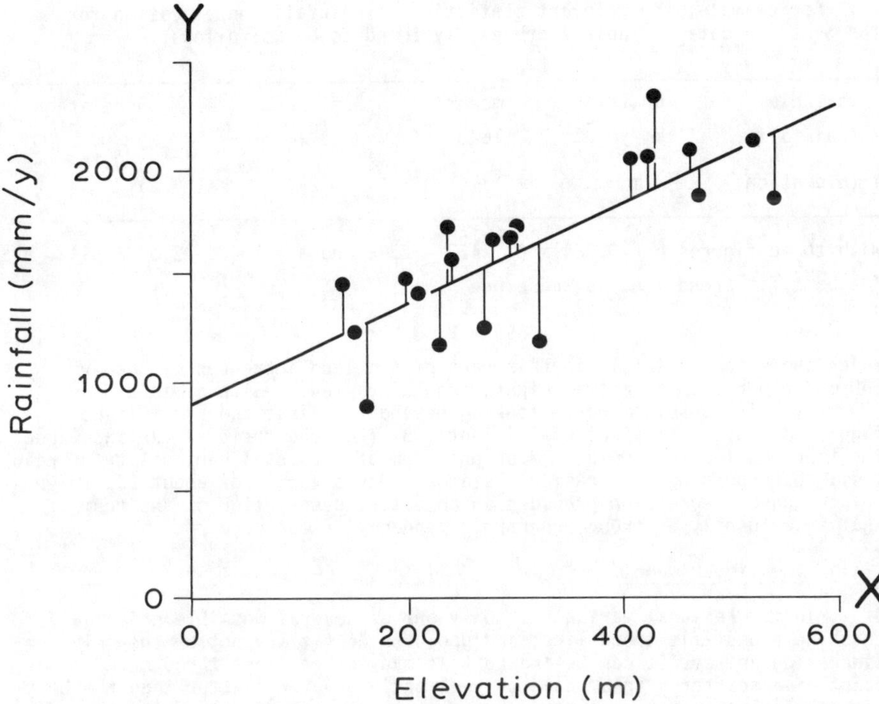

Fig. 2 Scatter diagram of rainfall against elevation for southern Scotland. The trend line shown minimises the residual variance

The amount of residual scatter, measured by its variance, is now

$$s_e^2 = s_y^2(1-r^2) \qquad (2c)$$

and this must be less than for any other trend line through the same data. The stronger the correlation between the variables the less the scatter about the best fit trend line, until with r = 1 or -1 there is no scatter at all and every data point lies exactly on the trend line. The correlation coefficient, or more precisely its square, is therefore a measure of the goodness of fit of a least squares simple regression.

Equations (2a) and (2b) are general formulae, called estimators, for deciding the intercept and slope of the best straight-line description of a simple trend. When fitting a trend line to new data, there is no need to go through the mathematics of minimising the scatter, one simply inserts the values of the appropriate means, standard deviations, and correlations in the formulae.

For example, the relevant statistics of rainfall and elevation for the Scottish data of Table 1 are easily found to be as follows.

variable	units	mean	st. dev.	correlation
X (rainfall)	mm/yr	1640	371	)
				) 0.784
Y (elevation)	m	314	122	)

With these figures b = 0.784 x 371/122 = 2.38 and a = 1640 - 2.38 x 314 = 895. The best-fit trend line is therefore

$$\hat{Y} = 895 + 2.38 X \qquad (3)$$

which shows that rainfall in this part of Scotland increases by nearly 240 mm/yr per 100m of extra height, from a sea level value of just under 900 mm/yr (it is seldom worth looking beyond the first two significant figures of any statistical coefficient). By (2c) the residual variance about the trend is $100(1 - (0.784)^2)\%$ or only 39% of the total variability of rainfall, corresponding to a residual standard deviation s_e of about 230 mm/yr. Least squares regression provides an objective description of the form of what is evidently a strong orographic tendency in our data.

(ii) The linear model

Minimum residual variance is only one of several possible criteria for choosing a best-fit trend line, and goodness of fit may not be the only consideration anyway. It can be argued that many other lines through the mean point of a scatter diagram have only slightly greater scatter than the best-fit regression, and that a round-number slope that generalises to other data is the most useful description (Ehrenberg 1975, chs. 7,14). A further difficulty is that unless the correlation is perfect (r = 1 or -1), when statistical methods are unnecessary since the points lie on a straight line, the least squares regressions of Y on X and X on Y give different trend lines. One minimises the variance of residuals in the Y direction, the other that of residuals in the X direction. Which are we to take? If the aim is simply to describe the relationship the ambiguity is embarrassing, and as Ehrenberg notes it is impossible for both regression lines to generalise to different data.

The least squares method can however be justified if we are prepared to accept one of two alternative statistical models for our data. In this context a 'model' is a set of assumptions about the underlying nature of the relationship that is supposed to have generated our sample data, and the models are statistical because they involve chance. The two models make different assumptions and are applicable in different circumstances but the distinction is often blurred or ignored completely in the geographical literature. The first and less useful one assumes that paired values of X and Y are random variates which jointly follow a bivariate normal probability distribution. For this reason it is generally called the joint or bivariate normal model, though Poole and O'Farrell (1971) refer to it as the random-X model. A set of X,Y pairs from this model should show an ellipse-shaped scatter in an X-Y plot, densest in the middle and elongated in proportion to the correlation coefficient which in this model is a parameter of the probability distribution. The least squares regressions of Y on X and X on Y can now be shown to give the best possible estimates of, respectively, the value of Y given X and that of X given Y (see for example Sprent, 1969, ch. 2). So the regression of Y on X is appropriate for prediction of Y from X, and vice versa. But this still does not tell us which to use to describe the relationship, and for this purpose it seems more sensible to find the equation of the major axis of the scatter ellipse, estimated by the bisector of the two regression lines (see Till, 1973).

The joint model may be applicable when X and Y are different measures of essentially the same thing, but it is technically invalid when X or Y or both depart appreciably from a normal distribution, and logically inappropriate when there is reason to think that one variable (conventionally Y) depends on or is in some way a function of the other (X). The <u>linear regression model</u> provides an alternative justification for fitting a <u>trend by least squares</u> in these circumstances. To the extent that we rationalise the world in cause-effect terms this is a more versatile approach to regression. It is also technically more flexible in that X need not be either random or normally distributed. Instead we could for example measure rainfall (Y) at preselected and evenly-spaced heights (X).

The linear model assumes that the explanatory variable or predictor X affects the dependent variable Y in a systematic fashion that is distorted by more or less random scatter. Three possible reasons for this scatter are errors of measurement, idiosyncracies of the individuals to which the data refer, and neglect of other relevant factors - i.e. the relationship would hold exactly if other things were equal, but they are not. If one or more of these applies, the underlying relationship must be obscured but to an extent that we can only guess. We may suspect that our data are imperfect, individuality exists, and relevant variables have been overlooked - indeed the geographer can seldom be confident on any of these counts - but the size and direction of the disturbances so introduced are unknown.

Without further information the only way ahead is to lump the three kinds of complications into a single unobservable variable ε that stands for all sources of variability in Y other than X. In the linear model ε is taken to be added on to an unknown straight-line relationship between Y and X: that is, the ith observed value of Y is

$$Y_i = \alpha + \beta X_i + \varepsilon_i \tag{4a}$$

where α, β are unknown constants. The difference between this disturbed systematic relationship and the more fundamentally probabilistic joint model is sketched in Fig. 3. To proceed further we must assume that the mean disturbance is zero, i.e.

$$\overline{\varepsilon} = 0 \tag{4b}$$

and that the disturbances are uncorrelated with X values, i.e.

$$\overline{\varepsilon x} = 0 \tag{4c}$$

where, as before, x denotes the deviation of X from its mean $\overline{X}$. Assumption (4b) implies that the disturbances do not have the general effect of raising or lowering the Y values overall, whilst (4c) implies the absence of any systematic association between positive or negative disturbances and high or low X values.

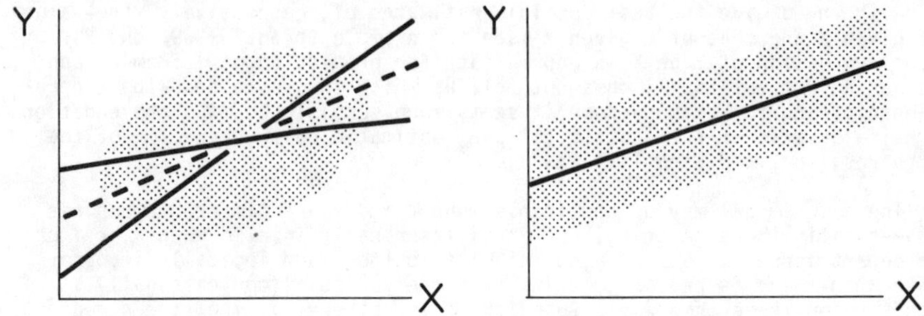

Fig. 3 Models for simple regression: (left) joint (right) linear. Solid lines are least squares trends

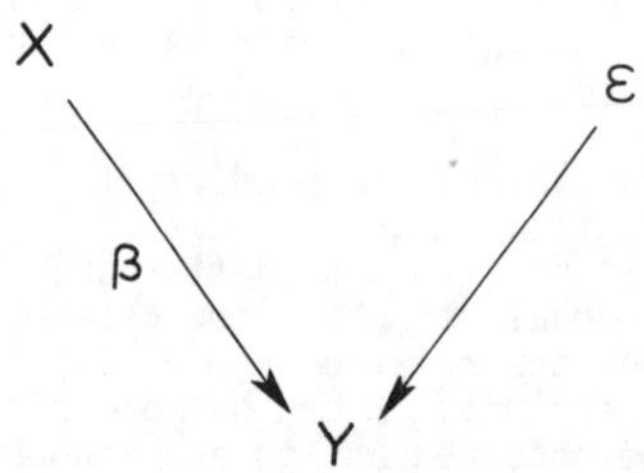

The situation can be represented diagrammatically by drawing arrows from causes to effects as shown. Note that changes in ε as well as in X lead to changes in Y.

The assumptions embodied in this model are sufficient to distinguish between the systematic and disturbance effects and thereby to identify the systematic relationship and the individual disturbances. First we eliminate the unknown intercept α. Since $\overline{\varepsilon} = 0$, averaging (4a) gives $\overline{Y} = \alpha + \beta\overline{X}$. This defines α once β is known:

$$\alpha = \overline{Y} - \beta\,\overline{X}. \tag{5a}$$

It can also be subtracted from (4a) to give the model in deviation form,

$$y = \beta x + \varepsilon.$$

Next, the covariance of X and Y can be looked at in two ways which must be equivalent. By definition it is $\overline{xy} = rs_x s_y$, which can be calculated from the observed data. But the model also tells us that $y = \beta x + \varepsilon$, so

$$rs_x s_y = \overline{xy} = \overline{x(\beta x + \varepsilon)} = \beta\overline{x^2} + \overline{x\varepsilon}.$$

By assumption (4c) the last term disappears, and $\overline{x^2} = s_x^2$, so

$$\beta = rs_y/s_x. \tag{5b}$$

The unknown coefficient β is thus determined by the standard deviations and correlation of the two observed variables, and α now follows from this and the observed means by (5a). Finally we can equate observed and model variances of Y:

$$s_y^2 = \overline{y^2} = \overline{(\beta x + \varepsilon)^2} = \beta^2\overline{x^2} + 2\beta\,\overline{x\varepsilon} + \overline{\varepsilon^2}$$

$$= \beta^2 s_x^2 \qquad + s^2_\varepsilon$$

or using (5b) and rearranging, $$s^2_\varepsilon = s_y^2(1-r^2). \qquad (5c)$$

The alert reader will have recognised equations (5a-c). Apart from the use of Greek letters for unobservable quantities, which is conventional, they are the same as the least squares formulae (2a-c). It follows also that the ε's and e's are identical. This overall equivalence between the two approaches is very important. It shows that the least squares method amounts to the assumption of a linear model with added zero mean disturbances uncorrelated with the systematic effect. Conversely the least squares trend identifies the true coefficients and disturbances of the underlying systematic relationship if these assumptions are correct. No probabilistic considerations are involved apart from the plausibility of the model.

The position is different if the observed data represent only a sample from some wider population of interest. This is the case in the Scottish example, where we have measurements for only twenty out of an infinite number of possible sites but would like to use the results to generalise about the regional climate or to predict rainfall at ungauged locations. The linear regression model (4) is now one possible view of the underlying relationship in the population as a whole. Let us assume it is appropriate. If the population means, standard deviations, and correlation were the same as the sample ones, then equations (5a-b) would give an accurate description of the population as well as the sample. But we have no means of telling whether this is so. The odds are overwhelmingly against, since the characteristics of a small sample usually differ from those of the parent population. Equation (5b) now gives only an estimate b (or $\hat{\beta}$) of β for the population, and consequently the intercept a (or $\hat{\alpha}$) given by (5a) is also likely to differ from the true value. In the same way the individual residuals e_1, e_2, ... will not be identical to the true disturbances ε_1, ε_2, ... Any predictions of ungauged rainfall will be doubly uncertain, with doubt about the correctness of the trend line added to the scatter of points around it.

Despite the inevitable uncertainty, detailed assessment of the probabilities involved shows that substitution of sample statistics in the assumed population model, or to put it a different way extrapolation of the sample least squares trend to the population, is a consistent method of estimation. It is the best possible method if the disturbance variable ε behaves in a simple random way to be discussed later. The routine mechanics of regression are therefore the same whether the aim is sample description, population inference, or prediction.

III MULTIPLE REGRESSION

(i) The need

The simple regression model just discussed assumes that differences in a dependent variable are accounted for partly by the linear effect of a single explanatory variable and partly by a disturbance term that lumps together individuality, measurement error, and the effects of relevant but unconsidered variables. It is not uncommon for the disturbance term ε to be numerically more important than the explanatory variable X, in the sense that the residual variance s^2_e is more than half the size of the variance of Y. Naturally we would like to do better than this and reduce the residual scatter as far as possible towards zero.

Nothing can be done about genuinely unique individual circumstances, and it may be impracticable to obtain more accurate measurements, so the main scope for improvement usually lies in separating out from the disturbance term other relevant explanatory variables and incorporating them explicitly in a multi-variable or multiple regression model. Most environmental systems involve a large number of interrelated variables so it is rare to find a satisfactory explanation for a spatial or historical pattern in terms of a single influence. Generally several conflicting or reinforcing effects are at work and the variables responsible are themselves interrelated. In such cases simple regression may be a very inadequate guide to the pattern of linkages and multiple regression is needed to clarify the situation.

The identification of potentially relevant variables is largely a matter of imagination and commonsense, guided by any relevant geographical theory. Several additional variables may be candidates for inclusion and in a later section we will see how to deal with them simultaneously, but for simplicity we start with just one. The first step is to obtain measurements of it for those individuals to which the original X and Y values referred. This is not always possible, so we may have to settle for some approximate indicator of the desired variable - a substitute or surrogate - and hope it behaves in essentially the same way.

Whether the new variable is indeed relevant can be tested by comparing the residuals from the original simple regression with corresponding values of the new third variable. For convenience let the original X become X_1 and call the new variable X_2. If a scatter diagram of residuals from the regression of Y on X_1, that is e's, against X_2 shows a definite trend then X_2 is relevant and we ought to consider it explicitly as a second predictor of Y. As an illustration, residuals from the regression of rainfall on elevation in southern Scotland are listed in Table 2 in the same west-east order as the original data of Table 1. It is clear that the residuals tend to be positive in the west, negative in the east. In other words elevation on its own tends to underestimate rainfall in the west but overestimate it in the east. If the residuals are plotted against distance east (Fig. 4) a pronounced and approximately linear trend is apparent. The effect of location can also be detected in the original scatter diagram of rainfall against elevation if points representing sites west and east of the overall mean location are distinguished (Fig. 5). The two sets of points are clearly staggered, rather than super-

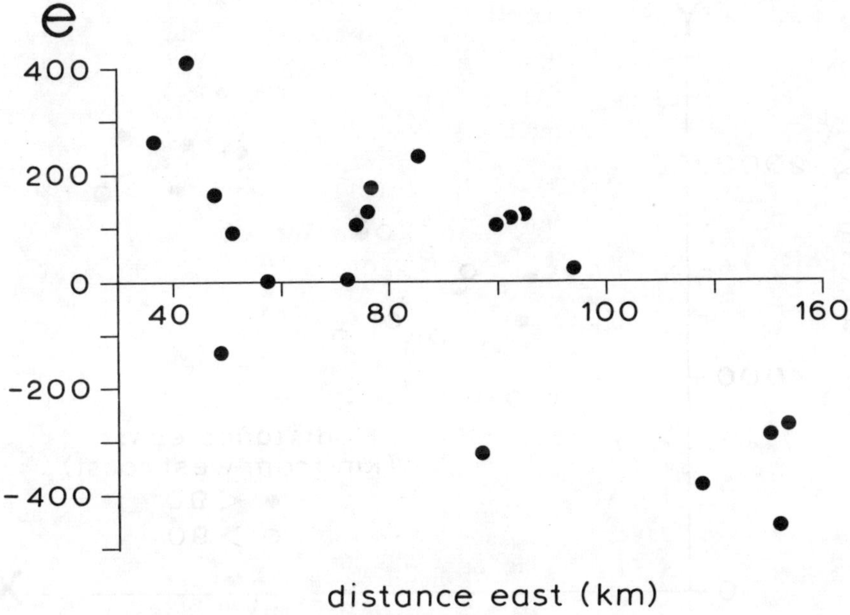

Fig. 4 West-east trend in residuals (e) from least squares regression of rainfall on elevation

imposed as would be expected if location made no difference once elevation had been taken into account. All this suggests an oceanic effect as well as the orographic one already considered.

How then do we include X_2 in the regression? One possible approach would be to fit a simple regression of the form $e = a + b X_2$ to the residual plot of Fig. 4 as an oceanic correction to be added on to the original regression equation (3). This is valid if X_1 and X_2 are uncorrelated either by chance or because of some deliberate sampling design, for example measurement of rainfall at a standard series of elevations at each of several different distances across the country. But it is rare for predictors to be completely uncorrelated when existing data are taken as they stand, and a sampling design is no help if some combinations of conditions are absent. In our Scottish example the correlation between elevation and distance east is not zero but -0.35, i.e. the high ground tends to be in the west, and this must be taken into account when attempting to separate the two variables' effects on rainfall. One way is to regress the residuals from the west-east trend of rainfall on those from the west-east trend of elevation, and those from the altitudinal trend of rainfall on those from the altitudinal trend of easterliness. This sounds, and is, complicated. Fortunately there is an easier method.

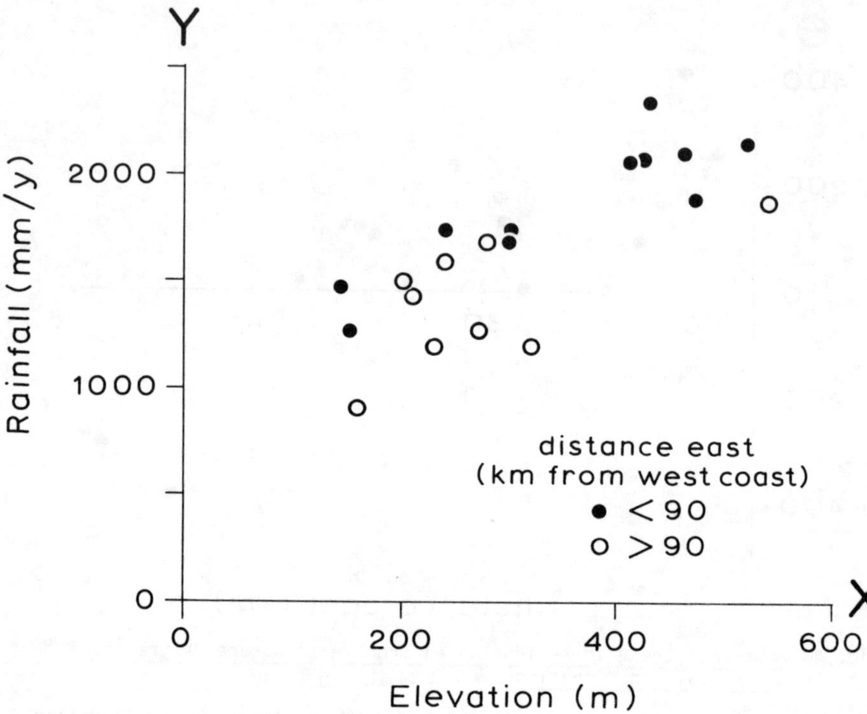

Fig. 5 Rainfall-elevation scatter diagram of Fig. 1 with western and eastern sites distinguished to show oceanic effect

(ii) Two explanatory variables

Multiple regression disentangles the effects of correlated explanatory variables after the event, statistically rather than experimentally. This is of course achieved at a price: we have to assume the type of relationship present, i.e. a statistical model for the data, and hope it is not so unrealistic that the results are misleading or meaningless.

The simplest and most commonly used model is linear and an extension of that for simple regression. The dependent variable is taken to be made up of a systematic effect plus a more or less random disturbance that averages out to zero and is uncorrelated with the systematic effect. The latter is assumed to be some linear function or weighted average of the two explanatory variables X_1 and X_2 (the term 'independent variables' should be avoided since they may well be correlated with each other and with other variables). Values of the dependent variable Y are therefore predicted by

$$\hat{Y} = a + b_1X_1 + b_2X_2 \qquad (6)$$

and are assumed to have been generated by the linear model

$$Y = \alpha + \beta_1X_1 + \beta_2X_2 + \varepsilon \qquad (7)$$

where α, β_1, β_2 are the unknown constants of the underlying systematic relationship and ε is the disturbance term. The intercept a or α has much the same meaning as in simple regression in that it fixes the level of Y when both X_1 and X_2 are equal to zero. But the b's or β's are not the same as simple regression coefficients of Y on X_1 and X_2. Instead, as can be seen from the fitted and model equations, b_1 or β_1 measures the effect on Y of a unit increase in X_1 with X_2 held constant, while b_2 or β_2 indicates the effect of a unit increase in X_2 with X_1 fixed. The 'other things being equal' proviso, analogous to the way an experimenter 'controls' all factors but one, is emphasised by calling the b's and β's partial regression coefficients. They are often distinguished from simple regression coefficients by a special subscript notation in which, for example, $b_{y1.2}$ represents the effect on Y (first subscript) of X_1 (second) with X_2 (after the dot) held constant. This is the same as b_1 in equation (6); b_2 there would be written as $b_{y2.1}$. In simple regressions of Y on X_1 and X_2 separately other things are not necessarily equal and the coefficients would be written simply as b_{y1} or b_{y2} without dots.

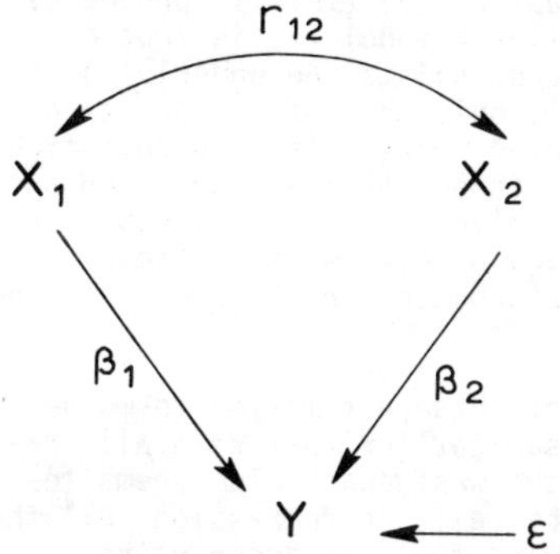

The assumed cause-effect structure is pictured alongside. The dependent variable Y is affected by changes in X_1, X_2, and ε, and X_1 and X_2 are themselves correlated as shown by the double-headed arrow. At this stage it does not matter whether their correlation reflects a causal link or if so which way round. As in simple regression the underlying systematic relationship can be identified by assuming the disturbances average out to zero ($\overline{\varepsilon} = 0$) and are uncorrelated with the systematic effects ($\overline{\varepsilon X_1} = \overline{\varepsilon X_2} = 0$). The first assumption means that the model can be averaged to give

$$\alpha = \overline{Y} - \beta_1\overline{X_1} - \beta_2\overline{X_2} \qquad (8a)$$

and consequently can be rewritten in deviation form as

$$y = \beta_1x_1 + \beta_2x_2 + \varepsilon \ .$$

The second assumption can now be used to simplify the results of equating observed covariances with those implied by the model:

$$r_{y_1}s_ys_1 = \overline{x_1y} = \overline{x_1(\beta_1x_1 + \beta_2x_2 + \varepsilon)}$$

$$= \beta_1s_1^2 + \beta_2r_{12}s_1s_2$$

and $$r_{y_2}s_ys_2 = \overline{x_2y} = \beta_1r_{12}s_1s_2 + \beta_2s_2^2 .$$

These two equations can be solved for the β's in terms of the correlations and standard deviations of the variables:

$$\beta_1 = \frac{s_y}{s_1} \frac{r_{y1} - r_{y2}r_{12}}{1 - r_{12}^2} \tag{8b}$$

$$\beta_2 = \frac{s_y}{s_2} \frac{r_{y2} - r_{y1}r_{12}}{1 - r_{12}^2} . \tag{8c}$$

The systematic part of the regression model can therefore be identified from simple descriptive statistics of the observed variables.

Before this is illustrated three features of these equations should be noted. First, they are symmetric: one can be obtained from the other by swapping subscripts 1 and 2. This reflects the equal status of X_1 and X_2 in the model and means in practice it does not matter which explanatory variable is labelled as X_1 and which as X_2. Second, if $r_{12} = 0$, (8b) and (8c) reduce to separate instances of formula (5b) for a simple regression coefficient. So if, and only if, the X's are uncorrelated their effects on Y are separate and are identified correctly by simple regression. Third and most important, expressions (8a-c) for α and the β's are identical to those for the least squares values of a, b_1, b_2 in the prediction equation (6) (this is proved by calculus in most advanced texts). Thus if the linear model (7) is true of the observed data the least squares trend is identical to the underlying systematic relationship. If however the model is only true of a population from which the data are a sample then the observed means, standard deviations, and correlations are unlikely to be the same as those for the population as a whole. Their use in equations (8a-c) now only gives estimates a, b_1, b_2 (or $\hat{\alpha}, \hat{\beta}_1, \hat{\beta}_2$) of the true coefficients. But it can again be shown that these estimates are the best possible ones if the disturbances or ε's behave in the simple random manner discussed later.

As an example of the routine application of these estimation formulae, consider again the Scottish rainfall data. We saw earlier that rainfall increases with elevation but that distance from the west coast also seems to be relevant. The two effects can be separated by multiple regression. All the necessary information is included in the following table of descriptive statistics calculated from the data of Tables 1 and 2.

variable	mean	st. dev.	correlation with Y	X_1	X_2
Y (rainfall)	1640	371	1		
X_1 (elevation)	314	122	.784	1	
X_2 (distance E)	89.4	36.7	-.730	-.353	1

Substitution of these figures in formulae (8a-c) gives partial regression coefficients $b_1 = 1.82$, $b_2 = -5.23$, and intercept $a = 1540$. The fitted regression equation is therefore

$$\hat{Y} = 1540 + 1.82X_1 - 5.23X_2 . \tag{9}$$

This can be interpreted in three ways. It is an objective description of the observed relationship between the three variables. It is moreover an accurate description of the underlying systematic relationship if the linear model (7) and associated assumptions are true for the 20 sites. Alternatively it is the best available estimate of the underlying relationship if the linear model is true of the region as a whole and the assumptions to be described later are true of the disturbances.

The estimated values of the coefficients show that average precipitation tends to increase with altitude by about 180 mm/yr per 100m at a given distance from the coast, and to decrease by over 5 mm/yr for every kilometre eastwards at any given height, from a base figure of 1540 mm/yr at sea level on the Ayrshire coast. The suggested rate of increase of rainfall with elevation is appreciably lower than in the simple regression (3) considered earlier, so taking location into account makes a difference to the apparent orographic effect as well as adding an oceanic term to the equation. We will come back to this point later on in discussing the various ways three variables can interrelate.

Just as the fitted simple regression could be represented by a sloping trend line in a two-dimensional scatter diagram, so the fitted multiple regression (8) can be seen as a tilted trend plane in a three-dimensional scatter of points (Fig. 6). The partial regression coefficients b_1, b_2 define the slope of the plane in the X_1 and X_2 directions respectively, while the intercept a determines where the plane cuts the Y axis. In this case b_1 is positive but b_2 negative, so the plane dips from back to front in Fig. 6: the maximum contrast in rainfall is between high ground in the west and low ground in the east, not simply west and east or high and low ground. These two situations would be represented instead by planes with no slope in one direction, i.e. $b_1 = 0$ or $b_2 = 0$. If rainfall did not depend systematically on either height or location the regression plane would be horizontal with both b's equal to zero.

Multiple regression in this case gives a more accurate description than simple regression of the regional distribution of precipitation, but there is of course still some residual scatter. In Fig. 6 the individual data points must be thought of as lying a distance e above or below the regression plane. The residuals above the plane balance those below it ($\bar{e} = 0$) and are not systematically associated with high or low values of either elevation or location (e is uncorrelated with X_1 and X_2). The scatter diagram of Fig. 5 can be interpreted as a view into Fig. 6 from the front right, with points differentiated according to their distance into the third or X_2 dimension. The fitted multiple regression could be added to this Y-X_1 diagram by drawing a series of parallel trend lines for different values of X_2. The rainfall-height trend is thus shifted up or down according to distance east. Clearly this ought to improve the overall goodness of fit, and we consider this next.

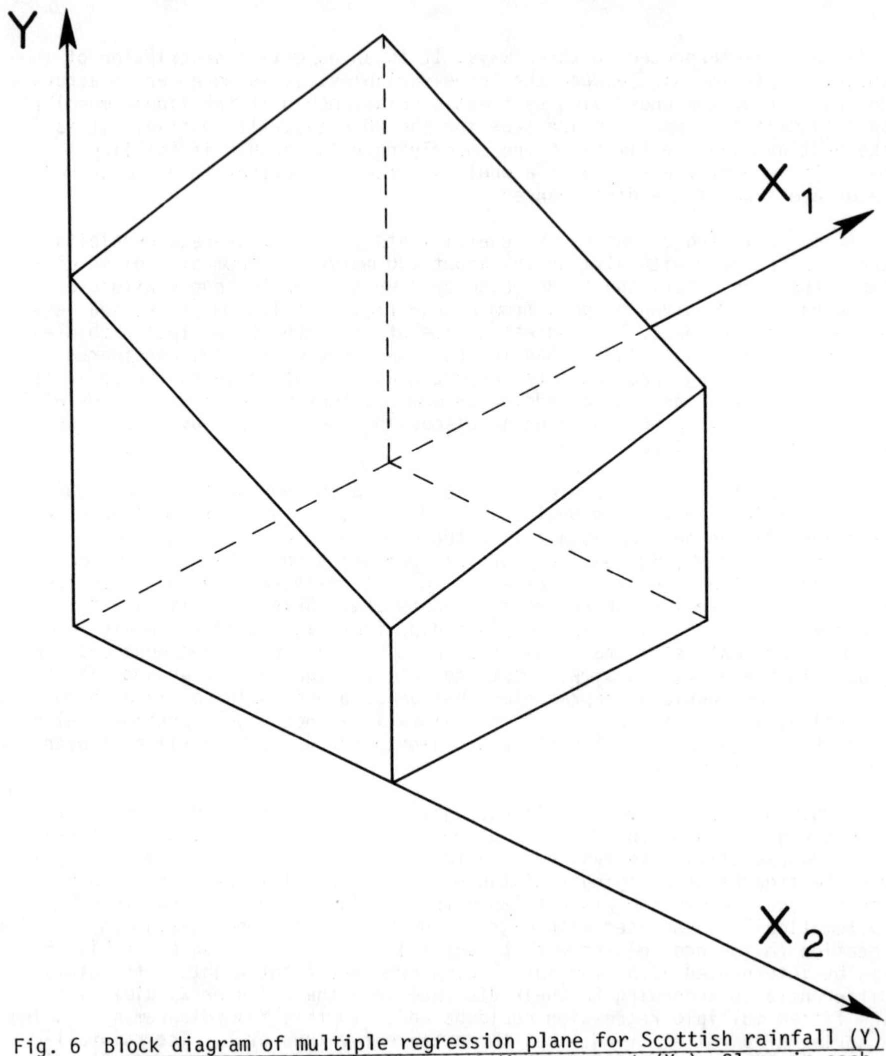

Fig. 6 Block diagram of multiple regression plane for Scottish rainfall (Y) as function of elevation (X_1) and distance east (X_2). Slope in each direction is proportional to partial regression coefficient

(iii) Multiple and partial correlation

The fundamental assumption of the linear regression model is that an observed imperfect trend would be a perfect systematic relationship but for more or less random disturbances. We have seen that the two components can only be separated by assuming that they are independent - that ε is uncorrelated with X in the simple regression model (4), or with both X_1 and X_2 in the multiple regression model (7).

It follows that the total scatter of the dependent variable Y is made up of two parts, that due to the systematic effect of the measured explanatory variables and that due to unsystematic disturbances:

total variance of Y = variance due to X's + variance due to ε.

The least squares method ensures that the residual term e is uncorrelated with the X's and therefore also with $\hat{Y}$, so essentially the same division applies to sample data:

observed variance of Y = variance of $\hat{Y}$ + variance of e

or

total variance = explained variance + unexplained variance.

'Explained' means of course 'numerically accounted for' and need not imply any understanding of why the relationship exists.

The relative sizes of the two variance components are obviously of great interest to anyone carrying out a regression analysis, whether the aim is prediction, hypothesis testing, or generalisation. The proportion of observed Y variance explained by a fitted regression equation is denoted by R^2 and called the coefficient of determination. The explained and unexplained proportions of the total variance are therefore

$$\left.\begin{aligned} R^2 &= \frac{\text{Y variance explained by regression}}{\text{total Y variance}} = s^2_{\hat{y}}/s^2_y \\ 1-R^2 &= \frac{\text{residual or unexplained variance}}{\text{total Y variance}} = s^2_e/s^2_y \end{aligned}\right\} \quad (10)$$

If R^2 is close to its maximum value of 1 the X's together account for a high proportion, $100R^2\%$, of the observed variability of Y and the residual scatter about the systematic trend is relatively low. If R^2 is close to its lower limit of zero the regression explains very little of the variability of Y and the residual scatter is large. In this way R^2 provides a dimensionless measure of goodness of fit, standardised to a 0-1 range. Its positive square root R, the multiple correlation coefficient, is the same as the simple correlation between the observed Y's and the $\hat{Y}$'s predicted by substituting observed X values in the fitted regression equation. In simple regression this is the same as the correlation of X and Y, so $R^2 = r^2$ and the definition (10) is simply equation (5c) in disguise.

It is not necessary to calculate individual residuals and find their standard deviation in order to evaluate R^2. Instead one can find the variance of $\hat{Y}$ by averaging over the observed data:

$$s^2_{\hat{y}} = \overline{\hat{y}^2} = \overline{(b_1x_1 + b_2x_2)^2} = b_1{}^2s_1{}^2 + 2b_1b_2r_{12}s_1s_2 + b_2{}^2s_2{}^2$$

which need only be divided by s^2_y to obtain R^2. The residual variance s^2_e can now be calculated as $s^2_y(1-R^2)$. With the data of the Scottish rainfall example $R^2 = 0.85$: that is, elevation and distance east together explain 85% of the observed variability in rainfall, leaving only 15% unexplained or residual variance attributable to unmeasured complications.

Figures like this are difficult to assess in isolation. They provide an objective measure of goodness of fit, but one's satisfaction with the result depends on temperament, past experience, and prior expectations. Since R^2 is defined as a proportion of the total variability of Y the nature of this base figure should also be kept in mind. A high coefficient of determination for a large set of data spanning a wide range of conditions - rainfall throughout Britain, perhaps - is more impressive than the same value of R^2 in a more restricted analysis.

It is also possible to make internal comparisons to see how far the second predictor improves the regression. For the simple regression of rainfall on elevation $r^2 = 0.61$, so taking distance east into account as well has lifted the level of explanation from 61% to 85%. This too must be assessed in comparative terms. Goodness of fit clearly cannot drop when an extra predictor is considered: even if the new variable is completely irrelevant R^2 will stay the same. At the other extreme the additional X may account completely for the previously unexplained scatter in Y, so that R^2 increases right up to 1 or 100%. If the actual improvement is expressed as a fraction of this maximum possible gain we have what might be called a coefficient of extra determination,

$$\frac{R^2\text{new} - R^2\text{old}}{1 - R^2\text{old}} = \frac{.85 - .61}{1 - .61} = \frac{.24}{.39} = .61$$

(the similarity to the original r^2 is coincidental). In this sense, then, taking location into account gives a 61% improvement in explanation compared to the simple regression of rainfall on elevation.

The square root of this quantity is called the <u>partial correlation</u> between the dependent variable and the new predictor with the old one (elevation) controlled or held constant. The partial correlation is always given the same sign as the corresponding partial regression coefficient, in this case b_2 of equation (9) which is negative, so we have

$$r_{y2.1} = \sqrt{.61} = -0.78$$

(the dot notation explained previously is used for partial correlations as well as regressions).

Multiple regression in this case provides a statistical substitute for the impractical experiment of flattening out southern Scotland to see more clearly the eastwards trend in rainfall at constant height. The variables can also be taken the other way round, starting with the simple correlation of -0.73 between rainfall and distance east. A similar calculation shows that the partial correlation between rainfall (Y) and elevation (X_1) controlled for distance east (X_2) is

$$r_{y1.2} = \sqrt{\frac{.85 - (-.73)^2}{1 - (-.73)^2}} = 0.82$$

which indicates that differences in height explain $(.82)^2$ or 68% of the rainfall variability not already accounted for by the simple west-east trend. In this way partial correlations supplement the partial b's as indications of the direct importance of individual explanatory variables.

(iv) <u>More than two predictors</u>

There is no logical limit to the number of variables that may be linked in a cause-effect web. There is however a statistical limit when regression analysis is used to establish the causal structure: if as many predictors are considered as there are individuals, we have one for each disturbance $\varepsilon_1, \varepsilon_2$, ... and might as well admit individuals are unique. In practice few geographers (or statisticians) would confidently interpret a regression on more than a few explanatory variables, so this indeterminate situation is unlikely to arise.

Regression on three predictors, then four, and so on could be explained by successively more complicated arguments of the same type as before, but it is neater and quicker to discuss the general case of p predictors. The prediction equation is

$$\hat{Y} = a + b_1 X_1 + b_2 X_2 + \ldots + b_p X_p \qquad (11)$$

which can be fitted by least squares or by considering the linear model

$$Y = \alpha + \beta_1 X_1 + \beta_2 X_2 + \ldots + \beta_p X_p + \varepsilon \,. \qquad (12)$$

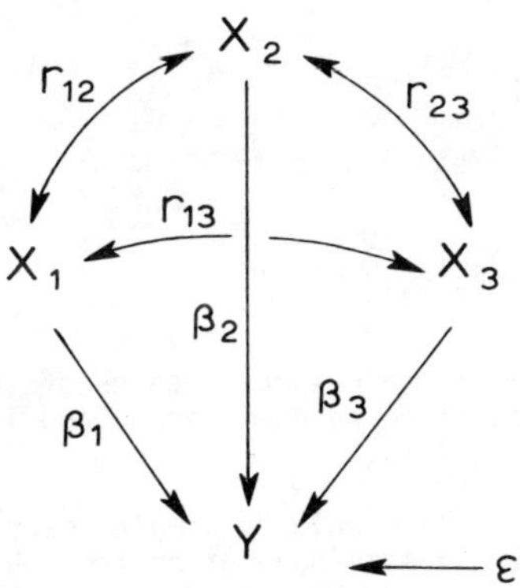

Values of the dependent variable are thus weighted averages of the corresponding values of the explanatory variables, plus a disturbance. The arrow diagram alongside shows the situation for p = 3 intercorrelated predictors.

To separate the systematic and disturbance effects we must assume disturbances cancel out overall ($\overline{\varepsilon} = 0$) and are not correlated with predictors ($\overline{\varepsilon x_j} = 0$ for each predictor $j = 1,2,\ldots,p$). The first assumption means that

$$\alpha = \overline{Y} - \sum_j \beta_j \overline{X_j}$$

and

$$y = \sum_j \beta_j x_j + \varepsilon \,.$$

We can now equate observed covariances with those implied by the model: for each predictor i = 1,2,...,p

$$r_{yi}s_ys_i = \overline{x_iy} = \overline{x_i\sum_j \beta_jx_j} + \overline{x_i\varepsilon}$$

$$= \sum_j \beta_j r_{ij}s_is_j$$

or $$r_{yi} = \sum_j (\beta_js_j/s_y)r_{ij} .$$

This set of equations for X-Y correlations in terms of β's and X-X correlations can be solved for the β's in terms of both sets of correlations. The more variables the longer the calculations, so a computer or desk calculator is normally used if more than two predictors are involved. Even so roundoff errors can lead to inaccurate results unless reliable equation-solving methods are used (see Mather and Openshaw, 1974).

The proportion of Y variance accounted for by the regression equation can also be found by an extension of the argument for two predictors. From equation (11)

$$R^2 = s^2_{\hat{y}}/s^2_y = \overline{(\sum_j b_jx_j)^2}/s^2y$$

$$= \sum_{ij}\sum(b_is_i/s_y)(b_js_j/s_y)r_{ij}$$

$$= \sum_i(b_is_i/s_y)r_{yi}.$$

The one- and two-predictor formulae given earlier are special cases of these general results. In all cases only the means, standard deviations, and inter-correlations of the observed variables are required.

Once again all this applies in the first place only to sample description. If the linear model is assumed to apply to some population from which our data are a sample, the descriptive statistics of the latter are likely to differ from those for the population and so therefore must the regression coefficients. But if the population disturbances (ε's) behave in a simple random fashion the b's obtained using sample data are the best possible estimates of the true β's. It is now time to examine more closely the extra assumptions involved.

(v) Statistical assumptions, residual checking, and transformation of variables

To ensure that regressions can be fitted to data we have already had to assume a certain type of relationship in which the dependent variable is made up of a weighted average of explanatory variables plus a disturbance. The weights or regression coefficients are constants for any one set of data, so according to the model a unit increase in one predictor changes Y by a fixed amount whatever the actual value of the predictor and irrespective of the values of other variables. These two properties are called linearity (since a plot of $\hat{Y}$ against any one X shows a straightline trend) and

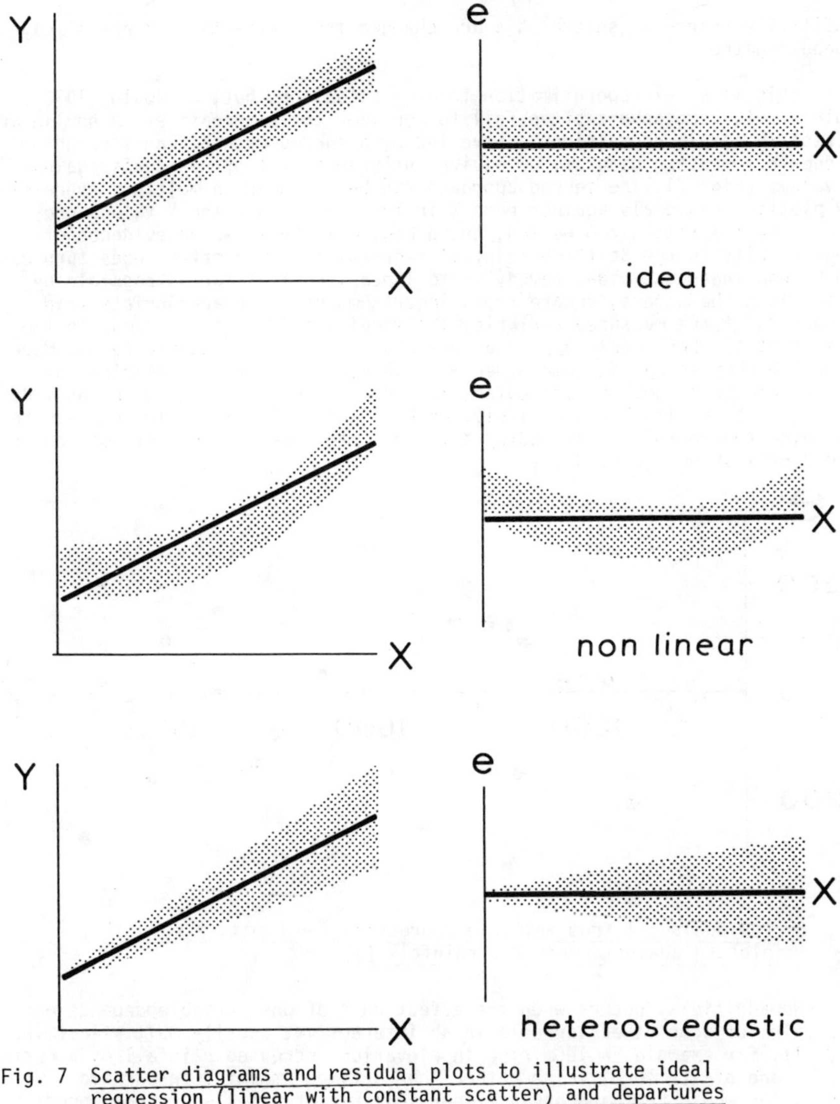

Fig. 7 Scatter diagrams and residual plots to illustrate ideal regression (linear with constant scatter) and departures from it

additivity (since if several X's are changed their effects on $\hat{Y}$ are simply added together).

This is a fair approximation to many situations but, as Gould (1970) points out, geographers often fail to consider the alternatives. A nonlinear effect in simple regression is revealed by a curved rather than straight trend in a scatter diagram, or equivalently in a plot of residuals against X values (Fig. 7). The second approach can be extended to multiple regression by plotting residuals against each X in turn or against the $\hat{Y}$ values predicted by the fitted regression, as in Fig. 8 which shows no evidence of nonlinearity in the Scottish rainfall regression. If an effect does turn out to be nonlinear the usual remedy is to linearise it as far as possible by using as X the square, square root, logarithm, or other appropriate transformation of the measured variable. The model should also be linear in the β's, that is they should appear as weights in a sum of separate terms. Models in which they appear in some other way can occasionally be linearised by transformations too. In particular, if Y depends on $e^{\beta X}$ or X^{β} then log Y depends on βX or βlog X respectively, both of which are linear in the required way even though the corresponding X-Y plots can take on a variety of concave and convex shapes (Fig. 9).

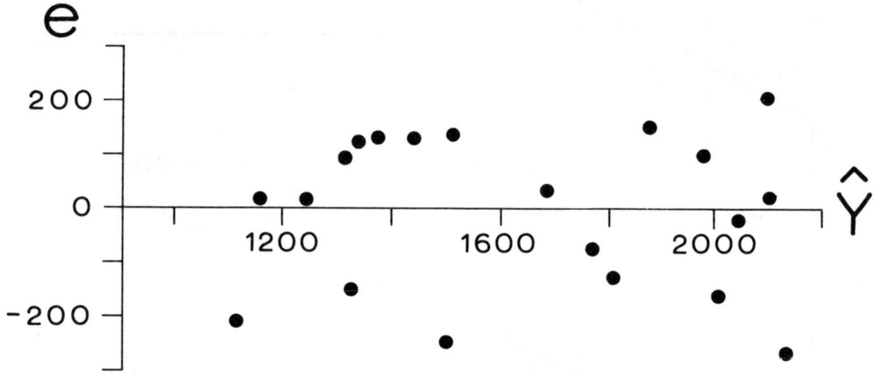

Fig. 8 Residuals (e) from multiple regression for Scottish rainfall, plotted against predicted rainfall ($\hat{Y}$)

Nonadditivity occurs when the effect on Y of one variable depends on the value of some other variable in an interactive, usually multiplicative, way. If, for example, a 100m rise in elevation increased rainfall by a certain percentage of its local value rather than a fixed amount then our additive regression would be inappropriate, particularly at sites where both predictors have extreme values. In the physical sciences multiplicative relationships are the rule, and dimensional considerations often indicate the appropriate form of combination (X_1X_2, $X_1/\sqrt{X_2}$, or whatever) which can then be used as a simple predictor (see Haynes, 1973 for a rare example of this kind of forethought in human geography). In less clear cut situations the most flexible approach is to take logarithms of all variables.

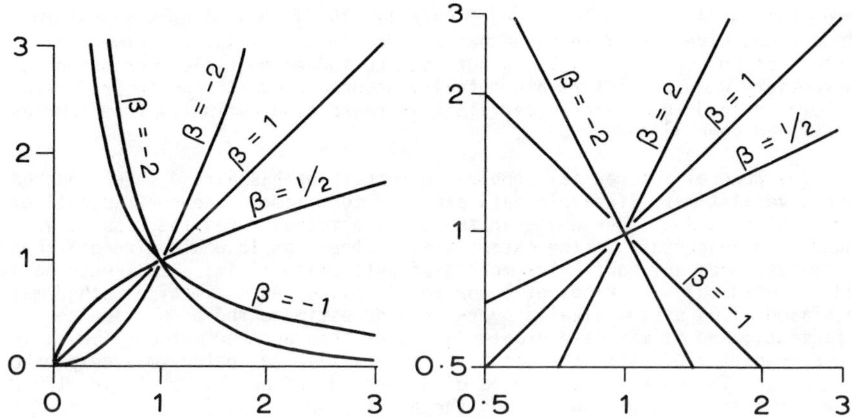

Fig. 9 <u>Arithmetic (left) and log-log (right) plots of power-law relationship $Y = X^\beta$ for different values of β</u>

This converts a multiplicative relationship such as

$$Y = 10^{\alpha} X_1^{\beta_1} X_2^{\beta_2} 10^{\varepsilon}$$

to the linear additive form

$$\log Y = \alpha + \beta_1 \log X_1 + \beta_2 \log X_2 + \varepsilon$$

which can be fitted by regressing log Y on the logs of X_1 and X_2 in the ordinary way. Note that the disturbance term in this is also interactive: its effect is to alter by a certain proportion or percentage, rather than absolute amount, the Y value determined by the systematic effects of the X's. We shall return to this topic in a moment.

Apart from linearity and additivity we have assumed so far that the disturbance term ε has zero mean and is uncorrelated with any predictor. These assumptions are not immediately testable since the least squares method ensures they are both satisfied, but the second one is related to the question of how many predictors to include in a regression. If we have omitted some relevant explanatory variable, say Z, that is correlated with one or more of the included X's, the true disturbance contains the 'lurking' variable Z and is not uncorrelated with each predictor. The assumption that it is leads to misleading estimates of the regression coefficients of those X's correlated with Z (Box, 1966; and see 'specification error' in econometrics texts). This problem does not arise if the omitted variable Z is uncorrelated with the included X's, or if it has no direct effect on Y, but the only way to be sure is to include potential lurking variables. They can always be abandoned if found to be irrelevant.

Another problem to do with correlations between variables is that if one X is completely predictable from one or more others, with $R^2 = 1$, then the variables concerned are said to be <u>multicollinear</u> and their effects cannot be separated. The least squares method breaks down in these circumstances and a regression cannot be fitted until the offending variable is omitted.

Some writers, including Poole and O'Farrell (1971) in an otherwise helpful discussion, give the impression that any non-zero correlation between predictors is unacceptable. This is not so, and indeed multiple regression is unnecessary when predictors are mutually uncorrelated. Strong intercorrelations do however lead to greater uncertainty in regression estimates from samples, as will be seen later.

The general inferential problem in regression has already been touched upon several times. If sample data are used to estimate the coefficients of a population model, then whatever the precise methods used the results are bound to be uncertain to the extent that another sample would give different estimates. Does any particular method of estimation minimise this uncertainty, and if so which? Two kinds of error are involved. An estimation method may be biased, i.e. systematically over- or underestimate the population coefficients; and it may have greater variance than another method, i.e. give a wider scatter of estimates about the true value. An analogy may help make this clear. The numbers at the top of a darts board are 5, 20, 1, 18. If 20 is the target, a player who throws three 1's shows bias; 5, 20, 1 is erratic, and 20, 1, 18 reveals both bias and variance. The best grouping is of course three 20's, with no bias and minimum variance.

If a linear additive model is an accurate description of a population relationship and the disturbance term ε behaves in a simple random fashion, least squares estimates of the regression coefficients using sample data are best linear unbiased estimates (BLUE for short). Proofs can be found in the advanced textbooks listed in the bibliography. The necessary assumptions about ε are that individual disturbances behave as uncorrelated random variables from probability distributions which all have a mean of zero and the same variance. In terms of expectations (averages over probability distributions),

(a) $E(\varepsilon_i) = 0$ for individuals $i = 1,2,\ldots,n$ and thus for all X values;

(b) $E(\varepsilon^2_i) = \text{constant} = s^2_\varepsilon$;

(c) $E(\varepsilon_i\varepsilon_j) = 0$ for all pairs of individuals $i \neq j$.

The first condition is a probabilistic analogy of our previous assumption that $\bar{\varepsilon} = 0$, and is reasonable if the linear additive model is appropriate or has been made so by variable transformations, and if no systematic 'lurking variable' has been left out. It is untestable unless there are several observations at each X value. This condition also ensures that ε is uncorrelated with any X, whether the latter has random or nonrandom (fixed) values, so long as these are measured without error. Inaccuracy in Y is permissible (indeed it was the original motive for developing regression methods), but errors in X's reduce their correlations with Y and necessarily lead to bias in the form of underestimation of the effect of each X (Mark & Peucker 1978).

The second condition, that of homoscedasticity, says that there is a constant degree of scatter about the population relationship rather than local regions of high or low scatter, when data points from high-scatter regions would exert undue influence on the least squares estimates. Nonconstant scatter or heteroscedasticity is commonly associated with disturbances that are proportional rather than additive. This is a special case of nonadditivity, as already discussed. It can be detected by inspection of the scatter diagram of a simple regression, or by plotting the residuals from a multiple regression against predicted $\hat{Y}$ values. Any systematic trend in the amount of scatter, as in the bottom graphs of Fig. 7, suggests heteroscedasticity. This time the multiple regression for Scottish rainfall does not pass the test so clearly, for there is some tendency towards greater scatter in the west where observed and predicted rainfalls are higher (Fig. 8). If this were more pronounced the least squares estimates would be less reliable. Proportional disturbances about a simple linear model can be accommodated by regressing Y/X on 1/X. Taking logarithms of all variables is another alternative, since as previously noted it converts multiplicative proportional disturbances to additive homoscedastic ones. More complicated types of heteroscedasticity can be dealt with only by weighting each observation: see 'generalised least squares' in advanced texts.

The third condition, that disturbances are mutually uncorrelated and therefore convey no information about each other, has been singled out as especially dubious by Gould (1970) and other geographers on the grounds that almost all geographical phenomena show positive spatial autocorrelation, with nearby places more alike. However, it is not necessary to assume that the values of any observed variable are mutually uncorrelated, only that the measurement error or extraneous complications affecting one Y value are unrelated to those affecting any other individual. Pronounced spatial patterns or trends in the X's or Y need not violate this assumption; what matters is that there should be no obvious pattern in maps or time series of residuals, or plots of residuals against $\hat{Y}$ or any X. There are several ways of testing this formally (see Cliff and Ord, 1972). Much the simplest is the runs test (described in most elementary statistics texts) in which the number of runs of successive residuals with the same sign in some appropriate plot is compared with the expected number for a random sequence. This is just over n/2, so the 20 residuals from the combined eastwards and altitudinal trend in Scottish rainfall pass the test comfortably with 11 runs in Fig. 8. Fewer and longer runs might have been found had raingauges within 2 km of others not been eliminated from the original sample, for local similarities in omitted variables such as aspect could lead to similar departures from the regional trend. Autocorrelation is therefore commoner in closely-spaced data. The same applies in time series, where temporary disturbances may carry over from one observation to the next if the interval is short. If substantial autocorrelation is present regression estimates may remain unbiased but no longer have minimum variance and are thus less reliable than usual. An intuitive explanation for this is that some of the data points are more or less duplicating each other so that the effective sample size is reduced. Unwin and Hepple (1974) discuss the problem further.

Visual inspection of residual plots generally provides the simplest and in many ways the best check of each of the assumptions that justify the use of linear least squares estimation. Residuals are, or can be, printed by most computer programs for regression analysis, and in some cases the plots too can be produced by machine, so residual checking is no great chore. It can however be less straightforward than suggested above if predictors have very skewed distributions. Scatter diagrams and residual plots then contain one or a few isolated points well clear of the rest, making it difficult to distinguish between trend and scatter. Log transformation helps here too by reducing positive skewness.

We have considered the statistical assumptions behind regression estimation at some length in order to see some of the problems that can arise in practice. Opinions differ as to how seriously violations of these assumptions should be taken. It is unfortunate that many geographical statistics books still in use stress only some of the assumptions, and then not necessarily the most important ones. Emphasis is often laid on the supposed need for normal distributions, but the distribution of each measured variable is irrelevant in linear regression and that of the residuals is relevant only to the significance tests discussed in the next section. Yet the important question of whether a multiplicative model with proportional disturbances is more appropriate than an additive one is seldom mentioned. It can also be argued that most geographical applications of regression analysis are exploratory, so that violations of the strict statistical assumptions are far less serious than in, say, the Treasury's predictions of economic variables such as unemployment and spending. As so often, the care needed with the tools depends on what is to be done with the finished product.

(vi) Confidence limits and significance tests

Regression coefficients obtained by the linear least squares method may be the best possible estimates of underlying relationships, but how uncertain are they? In particular, is the calculated b for a set of data sufficiently close to some hoped-for population β for the discrepancy to be only a matter of unlucky sampling? And how accurate are predictions made using the regression equation likely to be?

These are questions about confidence limits and significance tests, and they are relevant whenever we want to infer something about a population relationship from sample data. Even when we have data for every administrative area in a region it is sometimes argued that the boundaries could have been drawn in an infinite number of alternative ways, so that inferential questions are still relevant (see Gudgin and Thornes, 1974).

If the linear model $Y = \alpha + \beta_1 X_1 + \ldots + \beta_p X_p + \varepsilon$ is an accurate description of the population relationship between Y and the predictors X_1 to X_p, and the disturbance ε behaves randomly as specified in the last section, then the least squares estimates of b_1 from all possible samples of size n average out at the true value β_1. This is the 'unbiased' part of the BLUE property, and it applies also to the other b's and to a. The variance $s^2_e = s^2_y(1-R^2)$ of the residuals from the fitted regression is however not an unbiased estimate of the true disturbance variance s^2_ε, since the true residuals are bound to be bigger than those about the least-scatter line. The best estimate is instead

$$\hat{s}^2_\varepsilon = \frac{n}{n-p-1} s^2_e .$$

The square root of this quantity is sometimes called the <u>standard error of estimate</u>. It is the standard deviation of the distribution of each random ε value, and thus that of possible Y values about the true regression line (Fig. 10).

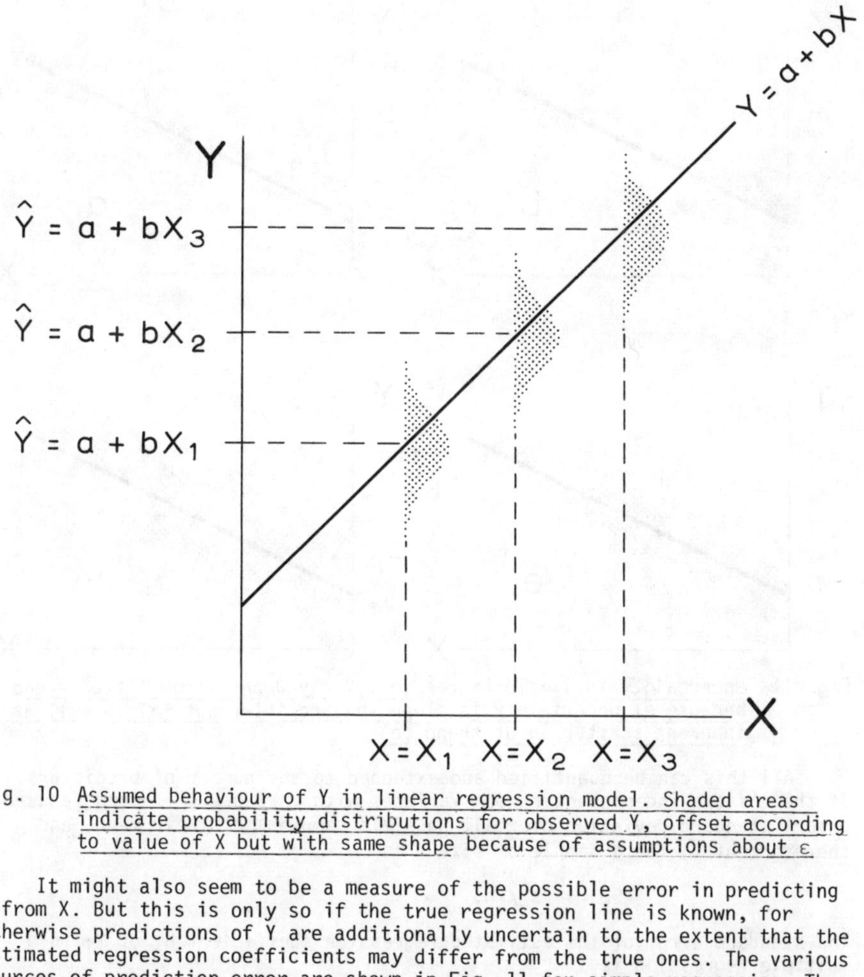

Fig. 10 Assumed behaviour of Y in linear regression model. Shaded areas indicate probability distributions for observed Y, offset according to value of X but with same shape because of assumptions about ε

It might also seem to be a measure of the possible error in predicting Y from X. But this is only so if the true regression line is known, for otherwise predictions of Y are additionally uncertain to the extent that the estimated regression coefficients may differ from the true ones. The various sources of prediction error are shown in Fig. 11 for simple regression. The shaded band in the lower left diagram indicates the range of Y values that can be expected simply because of variations in ε. But as the upper diagrams show the true slope may be steeper or gentler than the estimated value, and the sample and population centres of gravity through which the trend passes may differ. The overall uncertainty in predicting Y is the sum of all these components, and as shown in the lower right diagram it increases away from the mean of the data. Extrapolation beyond the range of the data is thus particularly uncertain, quite apart from the possibility that the relationship is not linear outside the observed range.

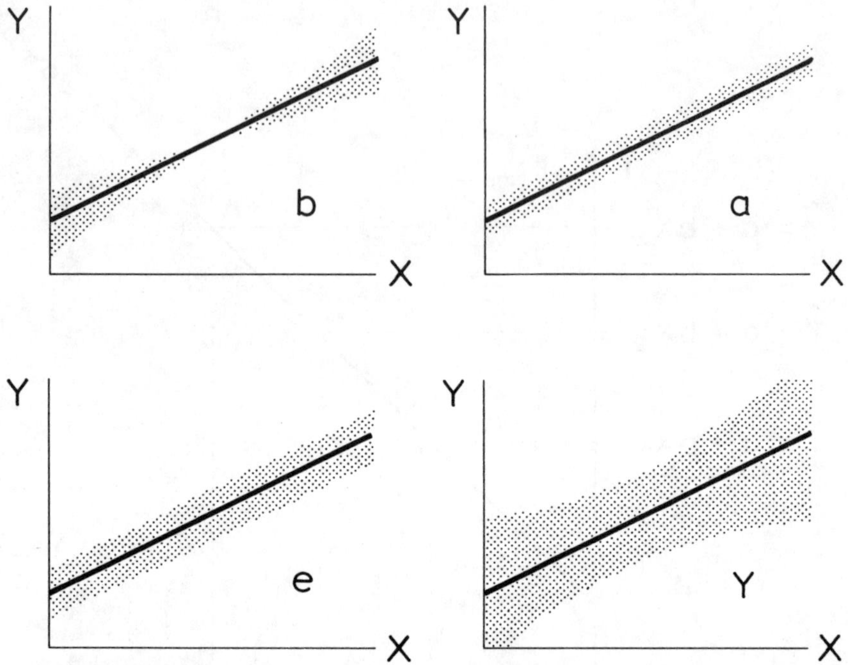

Fig. 11 Uncertainty in simple regression. Y may depart from fitted trend because of uncertainty in slope (b) and intercept (a) as well as inherent scatter about trend (e)

All this can be quantified and extended to any number of predictors. If the ε's are independent random variates with variance s^2_ε the standard error (standard deviation over all possible samples) of their mean, and therefore of $\overline{Y}$, for a sample of size n is

$$s_{\overline{Y}} = s_\varepsilon / \sqrt{n}.$$

The standard error of the estimated regression coefficient b_i of X_i in a multiple regression is

$$s_{b_i} = \frac{s_\varepsilon}{s_i} \sqrt{\frac{C_i}{n}}$$

where s_i is the standard deviation of X_i and C_i depends on the inter-correlations of the predictors. In simple regression C = 1, and with two X's $C = 1/(1-r_{12}^2)$ for each. Evidently regression estimates are most accurate when there is a wide range of X values but a small residual scatter and minimum intercorrelation of the predictors. In the multi-collinear case when two X's are perfectly correlated the standard errors of their b's are infinite. Given the standard errors of $\overline{Y}$ and each b, that of the trend value $\hat{Y}$ for any particular combination of X values can also be found. The standard error of the intercept a is a special case of this (all X's = 0) and the standard error of individual Y's can be found by adding on s_ε (see Draper and Smith, 1966, or any advanced econometrics text).

Without further information these standard errors can only be interpreted in relative terms. They all depend on the square root of n or something close to n, so a fourfold increase in sample size halves the uncertainty of regression estimates if other things are equal. But unless the probability distribution of ε is known we can only say within broad limits what proportion of samples are likely to give estimates within, say, two standard errors of the true value. The most convenient, and therefore commonest, assumption is that the ε's are independent random variates from the same zero-mean normal distribution. This makes the least squares estimates of regression coefficients not just BLUE but also maximum likelihood estimates: they are the values that maximise the overall probability of the sample data given the population model. The assumption of normal disturbances implies nothing about the frequency distribution of Y or of any X, so histograms of the measured variables are completely irrelevant. The probability distribution of Y conditional on the value of X is normal, but this is a different matter (Fig. 10). The only relevant test is to compare the histogram of calculated residuals against the normal curve with mean zero and standard deviation s_e, by eye or using the Kolmogorov-Smirnov test described in most elementary statistics texts. Fig. 12 illustrates this for the Scottish rainfall regression.

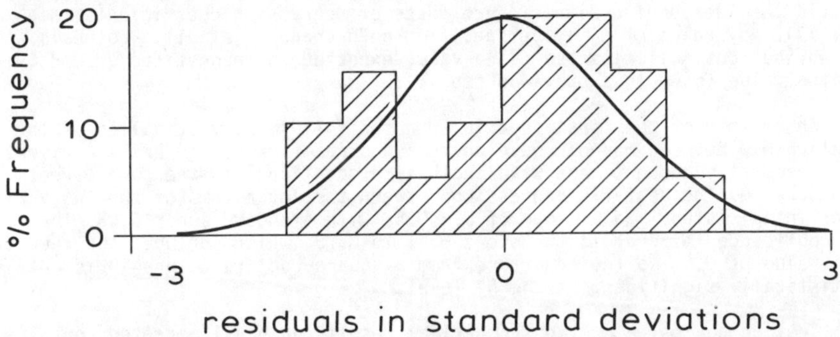

Fig. 12 Histogram of residuals from rainfall multiple regression compared to unit normal curve. Difference is not significant at 95% level by one-sample Kolmogorov-Smirnov test

If the ε's are normal variates the sample b's, a, $\bar{Y}$, and $\hat{Y}$'s, when divided by their estimated standard errors, all follow Student's t distribution with n-p-1 degrees of freedom (n-1 for $\bar{Y}$). Confidence limits can now be attached to estimates by multiplying the appropriate standard error by the tabulated t value for the desired confidence level, say 95% for which t is close to 2 except for very small samples. The true value of a quantity estimated from a sample regression is therefore about 95% certain to lie within 2 standard errors either side of the estimated value. Heteroscedasticity or autocorrelation in the disturbances increases the uncertainty and can make calculated confidence limits dangerously misleading (see Unwin and Hepple, 1974), but the assumption of normality appears to be less critical (Gudgin and Thornes, 1974).

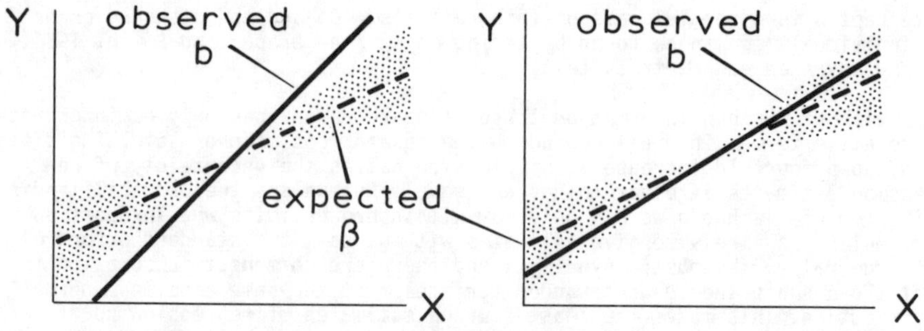

Fig. 13 Significant (left) and nonsignificant (right) differences between an observed sample b and an expected population β. Shaded area represents range within which b should lie with 95% or other chosen probability for samples from a population with the expected β

If the 95% (or 99%, or other) confidence limits for an estimated regression coefficient b lie on opposite sides of some prespecified value then we cannot safely say the true coefficient β differs from this value, since at least 5% (or 1%, etc.) of all possible samples from a population with the hypothetical β would yield at least as big a discrepancy as has been observed. This is the idea behind significance tests of regression coefficients (see Fig. 13). Any value of β may be specified beforehand as a null hypothesis. The obvious possibilities are (1) a value expected on theoretical grounds; (2) the value found in a previous study; (3) zero.

An example of the first type is Ferguson's (1975) investigation of the relationship between meander wavelength and streamflow for 19 British rivers. It is generally thought that wavelength is proportional to the square root of discharge. The best-fit regression between the logarithms of the two variables in this study had a slope of b = 0.58 with standard error 0.08. The 95% confidence interval is therefore 0.41 to 0.75, which includes the theoretical value of 0.5, so the departure from a square-root relationship is not statistically significant at the 5% level.

Testing against a previous empirical result can be illustrated for our best-fit relationship Y = 895 + 2.38 X (equation 3 above) between rainfall (Y) and height (X) at 20 sites in Scotland. This compares well with Y = 714 + 2.42 X found by Bleasdale and Chan (1972) for over 6500 rain gauges throughout the U.K. The standard error of b in the Scottish study turns out to be 0.44, giving a 95% confidence interval of 1.46 to 3.30 which easily includes Bleasdale and Chan's value of 2.42, even allowing for the latter's own standard error (which is very small because of the huge sample size). The orographic tendency in southern Scotland is therefore not significantly different from that for the whole U.K.

The third type of test, against the null hypothesis that β = 0, is even easier. It boils down to asking how many standard errors away from zero the observed slope, b, lies. Most computer programs for regression print the ratio of b to its standard error, generally labelled as 'T', and this can be compared with tables of Student's t. Our multiple regression equation (9) for Scottish rainfall has b_1 = 1.82 and b_2 = -5.23, with standard errors 0.31 and 1.02, giving t values of 6.0 and 5.1 (the sign is immaterial). These are overwhelmingly significant even at the 0.1% level. In other words the chances are far less than 1 in 1000 that we have simply an unrepresentative sample from a regional rainfall distribution that shows no systematic dependence on elevation or location ($\beta_1 = \beta_2 = 0$). It is however very unlikely that anyone would have entertained this null hypothesis seriously, so the significance test does not tell us much. It can be argued that tests of $\beta = 0$ are really only useful when there are strong grounds for expecting a particular effect to be absent or negligible, which of course takes us back to type (1) or (2) null hypotheses. And if the sample size is big enough even a tiny and geographically unimportant difference between sample b and expected β will be statistically significant.

The t test of $\beta = 0$ is mathematically equivalent to an analysis of variance or F test of the improvement in R^2 when the predictor concerned is brought into the regression. The variance ratio

$$\frac{(R^2 - R^2_{old})/(p - p_{old})}{(1 - R^2)/(n-p-1)}$$

can be used to compare any two linear regressions fitted to the same data and dependent variable. The numerator is proportional to the extra Y variance explained per new predictor, the denominator to that still unexplained by p predictors. The greater the improvement in R^2 and the smaller the residual variance the larger the ratio becomes and the less likely it is that the improved fit is a sampling fluke. If the ratio exceeds the tabulated value of the F distribution with $p - p_{old}$ and n-p-1 degrees of freedom, for some preferred significance level such as 5%, the contribution of the extra predictor(s) is statistically significant at this level.

Improvement in R^2 provides the definition of partial correlation, and in fact for a single extra predictor ($p - p_{old} = 1$) the variance ratio reduces to

$$r_.^2(n-p-1)/(1-r_.^2)$$

where $r_.$ is the partial correlation between Y and the new X with previous X's held constant. The stronger the partial correlation the more significant the effect of the new predictor. This test gives identical results to the t test of $\beta = 0$ so there is no need to do both.

There is nothing to stop us applying this F test to the improvement when several predictors are added to the regression, or to the improvement over no predictors at all. We can therefore test the overall goodness of fit of a multiple regression by setting $R^2_{old} = p_{old} = 0$ (in simple regression, i.e. p = 1, this is of course identical to the solitary 'partial' test). Most regression programs print out the overall variance ratio

$$\frac{R^2(n-p-1)}{p(1 - R^2)}$$

for comparison with the appropriate tabulated value of F. R^2 has to be very small not to be significant (less than 0.2 for large samples at the 5% level). If so the t or F tests of individual predictors are usually non-significant too. But this does not always apply in reverse: the regression as a whole may be significant, but not any of the individual effects. This paradoxical result occurs when a pair or set of predictors are so highly intercorrelated that controlling any one of them reduces their combined explanatory power to an insignificant level. This is yet another manifestation of the multicollinearity problem.

Partial t or F tests are often used as a means of screening variables and selecting the 'best' regression with a given number of predictors out of a wide choice (Draper and Smith, 1966, ch. 6). One way to do this is to start with the X having the strongest simple effect on Y (highest r^2) and try all multiple regressions of Y on this X and one other. The strongest of these (highest R^2, and thus including the new X with the strongest partial b or r) is now taken as a base on top of which all possible third predictors are tried. The process stops when the improvement on adding even the strongest extra X fails to reach some preset significance level. Alternatively one can work backwards and eliminate at each stage the X with the weakest partial b or r. A variant known as stepwise regression combines both approaches, working forwards but checking after each step to see whether any X has lost significance and should be dropped.

These selection methods are readily available in package programs and can save much time and effort. But for this very reason they are all too often a mechanical substitute for critical judgment about which variables ought to be relevant or irrelevant in the light of theory or experience. Automatic selection is also statistically suspect. It sometimes fails to find the regression with the highest R^2, and the sequential t or F tests involved are technically invalid because they are not independent (see Mather and Openshaw, 1974). The statistical assumptions about ε, violations of which generally make effects seem more significant than they really are, are rarely checked when large numbers of regressions are tried and discarded. And, most serious of all, the more significance tests are carried out the greater the chance of capitalising on a statistically significant but inexplicable result that really is a 1 in 20 or 1 in 100 sampling fluke. It must also be remembered that the tighter we set our significance levels to avoid accepting fortuitously strong effects, the more we are liable to reject real population relationships that happen to be weak in our sample data. This is by far the greater danger when samples are small. For all these reasons significance tests are a poor substitute for prior knowledge and critical judgment.

IV SPECIAL APPLICATIONS

(i) Causal models

Multiple regression analysis can be carried out for various reasons. One rather narrow application is the search for the best possible empirical equation for predicting one variable from a reasonably small set of others that are relatively easy or cheap to measure. Another is the estimation of a particular partial regression coefficient, i.e. the effect of one variable on another when extraneous complications are held constant, as a substitute for a controlled experiment or to substantiate earlier findings.

In much if not most social science, however, and much environmental science too despite its links with the 'exact' sciences, research is still exploratory. There is no consensus of opinion on which variables are relevant to which others, or how they compare in importance. Attention is focused on the existence and relative strength of relationships, not their precise form for which there are no empirical or theoretical yardsticks. This is a broader application of regression analysis, and involves assessment of the realism of alternative cause-effect models rather than calibration of one whose applicability is not in question. As an introduction to this kind of exploratory work we consider here the types of interrelationship that may exist between three variables, and only outline the extension to more complicated situations.

It will generally be clear which of three variables could in principle depend on both the others. This is Y and the others are X_1, X_2. Different situations can be recognised according to whether neither, one, or both of the X's does directly affect Y and whether they are themselves related. The cause-effect linkages can be represented by an arrow diagram. If either X has no direct systematic effect on Y the corresponding arrow can be omitted and the partial regression and correlation coefficients b and r. between Y and this X with the other X fixed must be equal to zero. If the X's are unrelated they do not need to be linked by an arrow and their simple correlation r_{12} and regressions b_{12}, b_{21} must equal zero. If r_{12} is not zero X_1 may affect X_2, be affected by it, or merely be correlated with it because both depend on some other variable not included in the analysis. These cases may be represented diagramatically by one-way and two-way arrows. If the three direct links are known the simple regressions between Y and each X can be found from the relationships

$$b_{y1} = b_{y1.2} + b_{y2.1}\, b_{21}$$

$$b_{y2} = b_{y2.1} + b_{y1.2}\, b_{12}$$

derived from the covariance equations which led to formulae (8a-b) for the partial regression coefficients. The following situations can be distinguished.

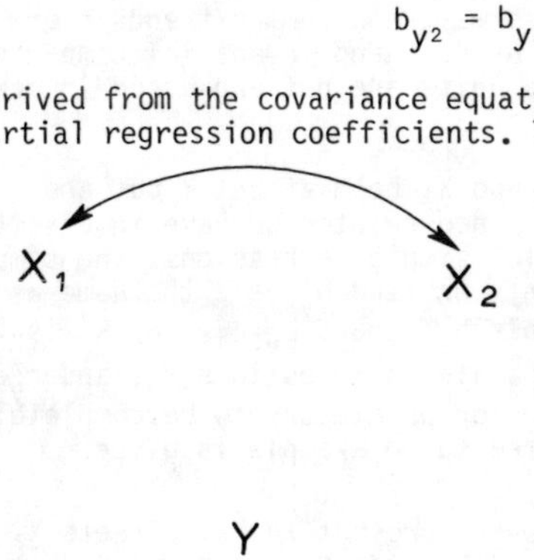

(1) If neither X_1 nor X_2 affects Y, all simple and partial correlations and regressions between Y and either X are zero and so is the multiple R^2. A correlation between the X's does not alter the situation.

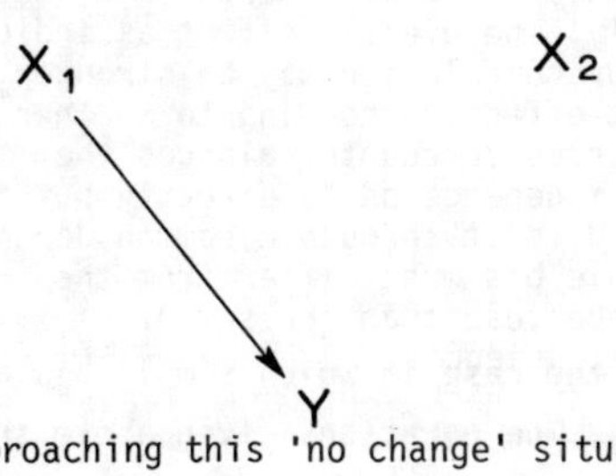

(2) The other situation with only one arrow is when one X has no effect on Y and is uncorrelated with the other X. It is then completely irrelevant and its partial and simple b's are both zero. The partial b for the other X is the same as the corresponding simple b, and $R^2 = r^2_{y1}$. Something approaching this 'no change' situation might be found if we controlled the

regression of rainfall on elevation for, say, the ages of the men who read the raingauges.

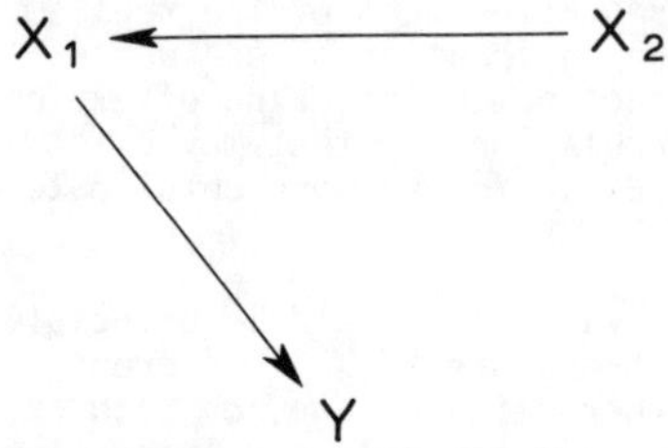

(3) One predictor, say X_2, may have no direct effect on Y but still have an indirect effect through X_1. For example, if rainfall (X_1) depends on elevation (X_2) and itself determines streamflow (Y), the latter will appear to increase with elevation even though little or no direct effect is involved. In this case $b_{y1} = b_{y1.2}$ and $R^2 = r^2_{y1}$, but b_{y2} is not zero like $b_{y2.1}$. Instead there is an apparent simple regression $b_{y2} = b_{y1}b_{12}$, with correlation $r_{y2} = r_{y1}r_{12}$.

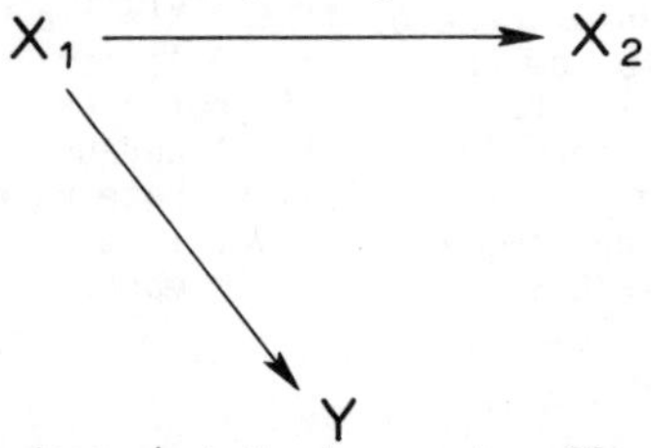

(4) A similar statistical situation arises if X_1 affects X_2 rather than the other way round, and X_2 still has no direct effect on Y. The common dependence on X_1 gives rise to a spurious correlation of strength $r_{y2} = r_{y1}r_{12}$ between Y and X_2, with a corresponding simple regression coefficient $b_{y2} = b_{y1}b_{12}$. Multiple regression again clarifies the situation by showing that X_1 does, but X_2 does not, affect Y directly. Spurious correlations are common in time series data, where pairs of variables may show common trends over time because of population growth or rising prices, and in spatial comparisons where variables measured for different-sized units are not expressed in per head or per unit area terms.

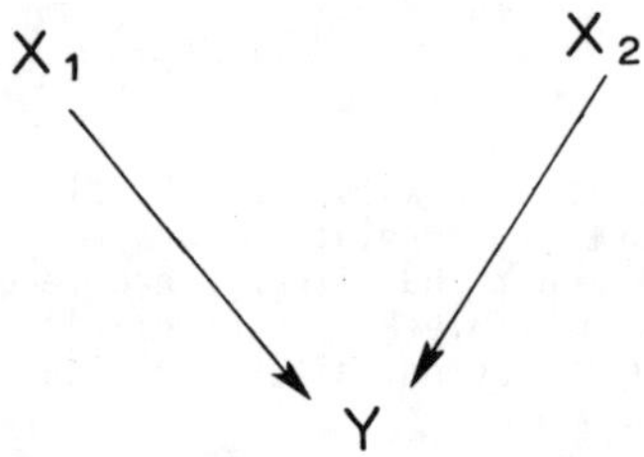

(5) If X_1 and X_2 both affect Y but are themselves uncorrelated we have in effect two parallel simple regressions. The simple coefficients b_{y1} and b_{y2} are the same as the partials $b_{y1.2}$ and $b_{y2.1}$, and R^2 is the sum of separate contributions r^2_{y1} and r^2_{y2}. It is rare for predictors to be completely uncorrelated so no example is given.

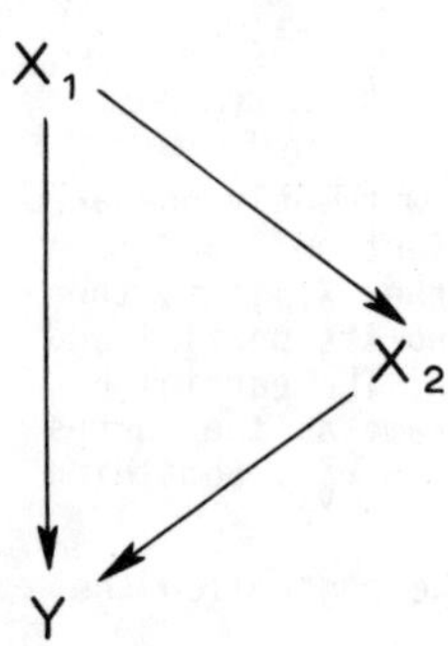

(6) If all three links are present and X_1 affects X_2 rather than the other way round, Y depends on X_1 both directly and indirectly. The overall effect as indicated by simple regression or correlation may be stronger or weaker than the direct effect, according to whether the indirect effect reinforces or counterbalances the direct one. Similarly Y depends on X_2 directly but is also spuriously related to it through a common dependence on X_1. Both simple b's must differ from the partials, and R^2 must be less than the sum of r^2_{y1} and r^2_{y2}. Reinforcement is the case in which simple b's are stronger than partials. The equations given above show

that if $b_{y2.1}$ is positive, b_{y2} exceeds it only if Y and X_2 both increase or both decrease when X_1 increases. Conversely if $b_{y2.1}$ is negative, b_{y2} is more negative only if Y and X_2 are affected in opposite directions by X_1. Thus none or two of the direct links must be negative for reinforcement to occur. Each predictor then amplifies the direct effect of the other. The Scottish rainfall example is a good illustration. If only simple regressions are considered the orographic effect is exaggerated by the oceanic influence also affecting most of the high ground, and vice versa. The modest correlation of -.35 between the predictors is sufficient to inflate partial b's of 1.8 and -5.2 to simple b's of 2.4 and -7.4 for height and distance east respectively. Whichever predictor is taken first, ignoring the other leads to a considerable bias in the regression estimate. Multiple regression is essential for a truer picture.

(7) The opposite case is suppression, where direct and indirect effects are in competition and tend to cancel out. This occurs whenever either just one or all three direct links are negative. What matters here is the sign of the partial (not simple) b describing the effect on Y of each X. A simple b can conceivably have the opposite sign to its partial if the indirect effect through the third variable outweighs the direct effect. It could even be zero if the direct and indirect effects cancelled out exactly (though this cannot happen to both simple regressions at once). Suppression would occur in the Scottish rainfall example if the topography of the region were reversed (r_{12} = .35 instead of -.35) without changing the prevailing air masses. There would now be only one negative direct link, that between rainfall and distance east. This would not affect the observed partial b's if the orographic and oceanic tendencies at work really are additive and linear. But the simple b's would drop to 1.3 and -3.1 and the corresponding simple correlations to 0.56 and -.41 instead of 0.78 and -.73 as actually observed. This is because orographic rainfall in the east would go some way to offsetting the oceanic influence in the west, giving a more uniform overall distribution of rainfall.

The different patterns of simple and partial b's, and to a lesser extent partial r's, that characterise these seven situations can be used to distinguish between alternative causal models for the observed relationship of three variables. This can be done by trial and error or systematically using the flow chart of Fig. 14. The diagnostic questions are whether neither, one, or both partials are so close to zero as to be negligible, and how the predictors are related. The first is a subjective matter when we only have sample data, for an effect may be absent in the population ($\beta = 0$) but present to some extent in the sample ($b \neq 0$). The significance tests outlined earlier give an indication of the likelihood that $\beta = 0$, and partials that are significant at better than 1% cannot lightly be ignored if the necessary assumptions are satisfied. On the other hand many investigators would not dismiss as negligible a partial that has the expected sign but fails to reach the chosen level of significance, since this could well reflect a sampling error of the second kind.

If both partials are adjudged negligible, neither X has an appreciable effect on Y; this is case (1) above. If one partial is negligible but the other not, the first X is either irrelevant (case 2, if it is also more or less uncorrelated with the other X) or only indirectly relevant (cases 3 or 4, according to the likely direction of the link between the X's; this too is a matter for the analyst's judgment). Finally if both X's are directly relevant the existence and sign of the correlation between them determines whether their effects are separate (case 5), reinforcing (case 6), or suppressing (case 7).

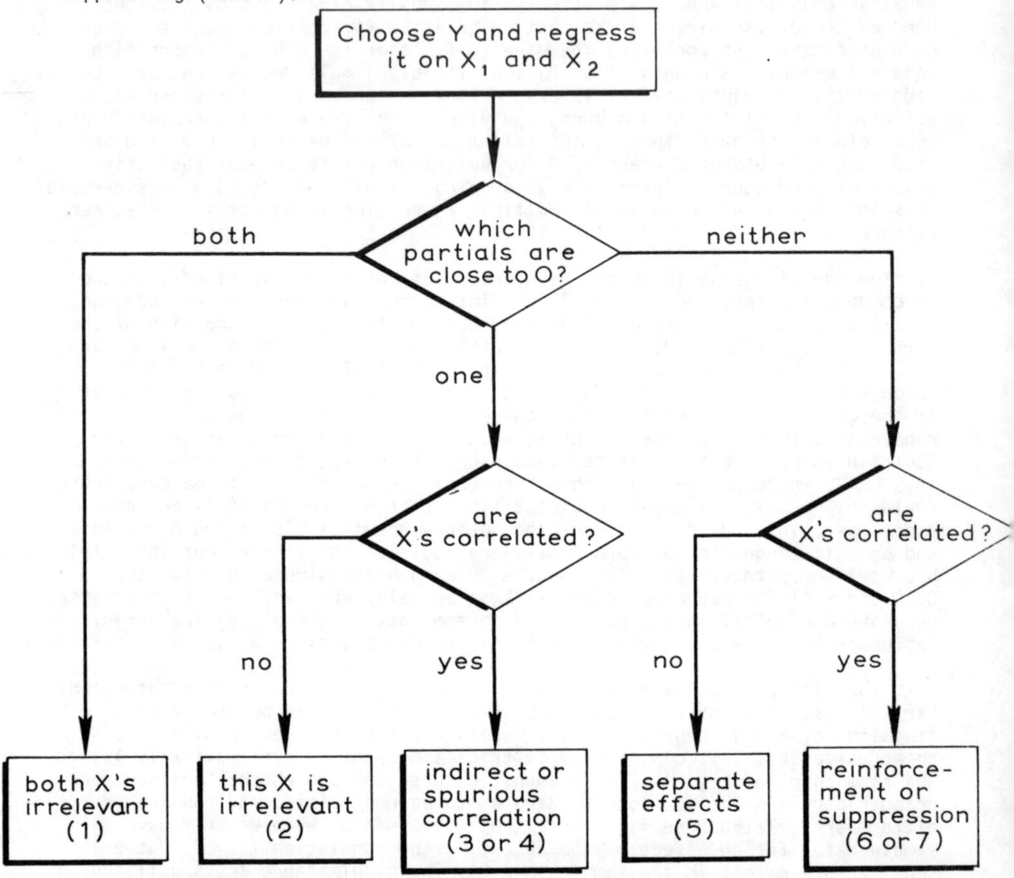

Fig. 14 Flowchart for diagnosis of three-variable causal models. Situations (1) to (7) are discussed in text

Causal models for relationships among more than three variables can be tested in a similar way if they can be represented by arrow diagrams without feedback loops. The general principle, first noted by the sociologist H.A. Simon, is that the absence of an arrow must be reflected in a near-zero partial b or r when any intervening variables and/or common causes are held constant. Repeated application of multiple regression or partial correlation to each partly or wholly dependent variable will either confirm the model or suggest necessary modifications. The classic text is that by Blalock (1961), and Mercer (1975) gives a clear account of an application in urban social geography. For what to do if feedback is present see Todd (1979).

The Simon-Blalock approach has much in common with the technique of path analysis, which originated in biology and is described in social statistics texts such as Heise (1975) and Kerlinger and Pedhazur (1973). The chief difference is that path analyses generally use standardised partial regression coefficients, which are b's measured in standard deviations of Y per standard deviation of X (they occurred unannounced in our discussion of regression on p predictors). The standardised form of a simple b is r, so the total effect of one variable on another is their correlation. It can be found from the arrow diagram as the sum, over all paths linking the variables, of the product of standardised b's along each path. In this way the importance of direct, indirect, and other paths between variables can be compared within one study. Standardised b's should not be used for comparisons between studies since they depend on sample variability, but the path analysis principle can be applied to unstandardised regressions as in our discussion of indirect, reinforcing, and suppressing effects.

(ii) Special kinds of variables

Regression analysis is mostly applied to variables that can potentially take any value within some continuous range. Rainfall, altitude, and distance from the coast must all be positive and have upper limits within any one region, but otherwise the number of possible values of each is restricted only by the imprecision of our measurements. This is not however true of all variables of interest to geographers. Some phenomena are simply present or absent, and others exist only in a few separate categories - types of house tenure, for example. These are examples of two-level and multi-level classifications. Classification is also commonly used for things that vary widely but are difficult to quantify on any continuous numerical scale: rock type, social class, and so on. At first sight relationships involving qualitative variables of this kind cannot be investigated by regression methods, but this is not necessarily so. We saw earlier on that the correlation coefficient r can be calculated for binary variables, i.e. those taking values 0 or 1 only. Since regression coefficients can be found from the correlations of predictors with each other and with a dependent variable, any qualitative phenomenon that can be expressed in binary form can be used in a regression analysis. Binary variables created for this purpose are called dummy variables. Their use enables several apparently different statistical techniques to be brought into the framework of regression theory.

One such application is the prediction of whether some qualitative phenomenon is present or absent, given values of one or more quantitative factors thought to be relevant. In this case we need a dummy dependent variable, say Y = 1 for presence or 0 for absence. If this is regressed on the X's in the usual way the fitted equation predicts presence if $\hat{Y}$ is greater than $\bar{Y}$, absence if $\hat{Y}$ is less than $\bar{Y}$. This technique is closely related to linear discriminant analysis (see King, 1969, 205-7).

The main drawback is that the scatter about the regression cannot be homoscedastic, so the least squares regression estimates are not as reliable as they could be. Wrigley (1976) in another monograph in this series describes a more sophisticated technique, logit analysis, that overcomes this problem and can be extended to qualitative dependent variables with more than two possible states. The basic idea is to regress not the dummy variable Y, but a transformation of it, on the X's. The predicted value $\hat{Y}$ can then be transformed back to get the best possible prediction of the probability of presence of whatever is represented by $Y = 1$. Wrigley gives the example of predicting how likely people are to suffer from acute bronchitis given their cigarette consumption and the CO_2 concentration in the atmosphere near their homes.

Dummy variables can also be used as predictors in a multiple regression model. If X_1 takes only the values 0 or 1, according to the presence or absence of some characteristic, the predicted value of Y for any given values of X_2, X_3, etc. is increased by an amount b_1 when the characteristic is present. In effect there are two separate regression equations which have the same coefficients for $X_2, X_3, \ldots$. but different intercepts, a when the characteristic is absent ($X_1 = 0$) and $a + b_1$ when it is present ($X_1 = 1$). For example, the magnitude of river floods is likely to increase with rainfall but may also depend on geology since the less permeable the ground the quicker storm rainfall gets into the river. Rock type can be taken into account by setting $X_1 = 0$ for permeable, 1 for impermeable, rock. Multiple regression of flood size (Y) on rock type (X_1) as well as rainfall (X_2) now gives two parallel trend lines in a plot of Y against X_2, both with slope b_2 but offset by a vertical distance b_1. The importance of geology can be assessed by the size and significance of the partial correlation coefficient $r_{y1.2}$, which will be larger the further apart the rainfall-runoff trend lines are for the two types of rock.

This method can be extended to several parallel trend lines, which amounts to analysis of covariance; to horizontal lines, which is equivalent to analysis of variance; to lines with different slopes but the same intercept; and to mixtures of all these cases. The multiple regression approach makes clear the links between the different possibilities, and allows easy comparison of their goodness of fit. Silk (1979) gives a detailed but readable account. Further applications of dummy predictors are described by Draper and Smith (1966, ch. 5) and Mather and Openshaw (1974).

Two further applications of multiple linear regression to special kinds of variables should also be noted. One is trend surface analysis in which Y is some spatially distributed variable and the X's are locational coordinates (e.g. eastings and northings) and their powers and products up to some maximum order. The aim is generally to see what order of surface adequately describes the geographical pattern of Y, using F tests of the overall R^2 and its improvement from one order to the next. Details and applications are described by Unwin (1975) in another monograph in this series.

The final special case is the autoregressive modelling of time series (Box and Jenkins, 1970). Here the X's are Y values one, two, etc. time intervals ago. The carryover effects in the series at these different time lags amount to partial regression coefficients of the series on its own past and can be found from the correlations between Y and the X's, i.e. the autocorrelations of the series with itself at different lags. Applications include the study of fluctuations in economic, climatic, and hydrologic time series. Spatial series in geomorphology have also been investigated in the same way, with distance (downslope or downriver) replacing time.

Multiple linear regression in its basic form is very widely used. The special applications to spatial patterns, time series, qualitative variables, and causal models make it even more versatile. The geographer who understands the fundamentals of linear regression is well placed to analyse most kinds of geographical data and to appreciate published quantitative research.

BIBLIOGRAPHY

A. Applications

Bleasdale, A. and Chan, Y.K., (1972), Orographic influences on the distribution of precipitation. 322-333 in: *Distribution of precipitation in mountainous areas*, 2, World Meteorological Office (Geneva).

Champion, A.G., (1972), Urban densities in England and Wales: the significance of three factors. *Area*, 4, 187-192.

Ferguson, R.I., (1975), Meander irregularity and wavelength estimation. *Journal of Hydrology*, 26, 315-333.

Haynes, R.M., (1973), Crime rates and city size in America. *Area*, 5, 162-165.

Krumbein, W.C., (1959), The sorting out of geological variables illustrated by regression analysis of factors controlling beach firmness. *Journal of Sedimentary Petrology*, 29, 575-587.

Mercer, J., (1975), Metropolitan housing quality and an application of causal modelling. *Geographical Analysis*, 7, 295-302.

Parker, A.J., (1974), An analysis of retail grocery price variations. *Area*, 6, 117-120.

Smith, G.C., (1976), The spatial information fields of urban consumers. *Transactions, Institute of British Geographers*, new series 1, 175-189.

Taafe, E.J., Morrill, R.L., and Gould, P.R., (1963), Transport expansion in underdeveloped countries: a comparative analysis. *Geographical Review*, 53, 503-529.

B. Assumptions and inference

Gould, P., (1970), Is *statistix inferens* the geographical name for a wild goose? *Economic Geography*, 46, 439-448.

Gudgin, G., and Thornes, J.B., (1974), Probability in geographic research: applications and problems. *The Statistician*, 23, 157-177.

Mather, P., and Openshaw, S., (1974), Multivariate methods and geographical data. *The Statistician*, 23, 283-308.

Poole, M.A., and O'Farrell, P.N., (1971), The assumptions of the linear regression model. *Transactions, Institute of British Geographers*, 52, 145-158.

Unwin, D.J., and Hepple, L.W., (1974), The statistical analysis of spatial series. *The Statistician*, 23, 211-227.

C. Advanced texts

Draper, N.R., and Smith, H., (1966), *Applied regression analysis*. (Wiley, New York).

Huang, D.S., (1970), *Regression and econometric methods*. (Wiley, New York).

Johnston, J., (1972), *Econometric methods* (2nd edition). (McGraw-Hill, New York).

Kerlinger, F.N., and Pedhazur, E.J., (1973), *Multiple regression in behavioral research.* (Holt, Rinehart and Winston, New York).

Surrey, M.J.C., (1974), *An introduction to econometrics.* (Clarendon Press, Oxford).

D. Other references

Blalock, H.M., (1961), *Causal inferences in nonexperimental research.* (University of North Carolina Press, Chapel Hill, North Carolina).

Box, G.E.P., (1966), Use and abuse of regression. *Technometrics,* 8, 625-9.

Box, G.E.P., and Jenkins, G.M., (1970), *Time series analysis, forecasting and control.* (Holden-Day, San Francisco).

Cliff, A.D., and Ord, J.K., (1972), Testing for spatial autocorrelation among regression residuals. *Geographical Analysis,* 4, 267-284.

Ehrenberg, A.S.C., (1975), *Data reduction.* (Wiley, London).

Heise, D.R., (1975), *Causal analysis.* (Wiley, London).

King, L.J., (1969), *Statistical analysis in geography.* (Prentice Hall, Englewood Cliffs, New Jersey).

Mark, D.M., and Peucker, T.K., (1978), Regression analysis and geographic models. *Canadian Geographer* 22, 51-64.

Silk, J., (1979), *Analysis of covariance and comparison of regression lines.* Concepts and techniques in modern geography, 20. (Geo Abstracts Ltd, Norwich).

Sprent, P., (1969), *Models in regression.* (Methuen, London).

Till, R., (1973), The use of linear regression in geomorphology. *Area,* 5 303-8.

Todd, D., (1979), *An introduction to the use of simultaneous-equation regression analysis in geography.* Concepts and techniques in modern geography, 21. (Geo Abstracts Ltd, Norwich).

Unwin, D.J., (1975), *An introduction to trend surface analysis.* Concepts and techniques in modern geography, 5. (Geo Abstracts Ltd, Norwich).

Wrigley, N., (1976), *An introduction to the use of logit models in geography.* Concepts and techniques in modern geography, 10. (Geo Abstracts Ltd, Norwich).